The Sulphur Cap in Maritime Supply Chains

Olli-Pekka Hilmola

The Sulphur Cap in Maritime Supply Chains

Environmental Regulations in European Logistics

Olli-Pekka Hilmola
Kouvola Unit
Lappeenranta University of
   Technology
Kouvola, Finland

ISBN 978-3-319-98544-2   ISBN 978-3-319-98545-9   (eBook)
https://doi.org/10.1007/978-3-319-98545-9

Library of Congress Control Number: 2018950555

This Palgrave Pivot imprint is published by the registered company Springer Nature
Switzerland AG
The registered company address is: Gewerbestrasse 11, 6330 Cham, Switzerland

Follow the little footsteps in the combat against climate change
and lower emissions –
oil producers and refiners are taking chips from the table.

PREFACE

I had a chance to live through the sulphur regulation change in Finland and Estonia during the years 2011–2015. In the very beginning, it was not even well recognized that regulation would be implemented at the demanding 0.1% maximum level; however, as the years progressed towards the implementation year of 2015, it became evident that nothing could stop this from happening. Many different parties (companies, sea ports, associations, NGOs, labour unions etc.) criticized this new legislation, and the most horrible scenarios indicated that a significant number of heavy industries and manufacturing would be lost (or severely hurt) due to this. The additional price tag for shipping costs for Finland alone was estimated to be several hundred million EUR annually (in some cases the price tag given was as high as more than one billion EUR p.a.). The price harm level in Sweden was argued to be at a similar level to Finland. Fortunately, all the most depressing scenarios were not realized, and companies responded to this change by trying to sustain their operations with the lowest possible cost impact (or harm). These small changes are so numerous that I saw it as important to write a book for the global audience in the face of the 2020 sulphur challenge.

The forthcoming change in 2020 is only a short time away, and many parties have not realized which is to come. What is comforting news is that the global sulphur level is set to a maximum of 0.5%. This is a demanding level, but yet at the same time it is a level that is much easier to manage than the maximum 0.1% content. For the 2020 sulphur cap, companies can still mix very low sulphur level diesel and higher grade diesel together to reach the acceptable level of 0.5%. In the year 2015 within the Baltic Sea and North Sea, this blending activity was basically out of the question as the sulphur level was set so low. However, what is demanding now is the tight implementation schedule (2020, instead of 2025), a big change from a 3.5% sulphur level down to 0.5%—and simultaneous global implementation. It is difficult to know how much implementation will eventually cost global maritime supply chains. Worldwide, it is definitely a cost tens of billions of EUR higher per annum, but it could even be 40–50 billion EUR.

Due to receiving an individual research grant for the year 2014, I was obligated to examine this transition from the Finnish perspective, and most of the findings of this book arise from this one-year period of intensive research on this topic. What has been fruitful in my own journey with the topic is the basic fundamental change that was already recognizable in 2012–2014. Very short sea shipping routes would benefit and attract volumes, and maritime supply chains would modify accordingly. This change has also had its implications on the amount of hinterland transport, and further dependency on high emitting road transportation. What was surprising in writing this book was to see that this was not only a matter for Finland, but that the also same applied to Sweden and Estonia. All three countries examined in this book (Estonia, Finland and Sweden) also experienced the fundamental change in their maritime transport system, where volumes were no longer necessarily growing in handled tons, but where maritime business was increasingly driven by unitized transport means. Heavy industries and mass manufacturing have swallowed the loss, but not suddenly; it was a rather gradual change (before and after 2015). In Finland, governmental support for maritime (fairway due payments of shipping companies, 50% reduction) and railway (track tax removal) transport during 2015–2017 (also continued in 2018) played a role in sustaining the volumes and foreign trade performance.

As so many parties are hurt and costs increase significantly, it is reasonable to ask which parties are benefitting from sulphur regulation tightening. The beneficiaries are society and its people, as the air becomes less polluted, which results in less sickness and fewer allergies, and even in fewer mortalities. This, of course, increases the quality of life and macro-productivity as well as lowering external costs in societies (typically paid for by the governments). It is clear after the experiences of 2015 that technology development and machinery sales related to decreasing shipping emissions could benefit from this change too.

In addition, oil refineries benefit very handsomely in financial terms. In the global implementation of 2020, oil refineries are likely to gather record profits out of supplying desperately needed low sulphur diesel oil with good profit margins and high volume. A friend in need is a friend indeed.

Kouvola, Finland   Olli-Pekka Hilmola

CONTENTS

CONTENTS

LIST OF FIGURES

LIST OF TABLES

CHAPTER 1

Introduction

Abstract Transportation logistics and supply chains in general are facing increasing environmental demands. It could be argued that the sector is coming late to the demands for reductions in greenhouse gases (GHGs) and environmental emissions. For a long time it was enough for global policies that trade and economies grew. This is no longer the case. The transportation sector is responsible for a significant amount of emissions, and in Europe, they are still on track for long-term growth. Most troubling among these modes of transportation is truck transportation. The main challenge in maritime supply chains is not the amount of CO_2 emissions, but that of sulphur and nitrogen emissions, and the concentration of these emissions to a limited amount of major sea ports.

Keywords GHG · Emissions · Trade · Transportation · Shipping

Globalization and trade growth have created the platform for actors in the logistics sector to grow and internationalize. Companies have become ever larger, and concentration to the only small number of global players, for example, in container shipping (United Nations 2017) and third-party logistics services (Bowman 2014), is already a fact. In addition, the market shares of the biggest companies are significant. These all changes have meant that margins of operation are becoming increasingly smaller, as competition is becoming more global and intensive.

© The Author(s) 2019
O.-P. Hilmola, *The Sulphur Cap in Maritime Supply Chains*,
https://doi.org/10.1007/978-3-319-98545-9_1

Nowadays, talk about environmental emissions and prevention of climate change is prevalent everywhere. Previously, factories controlled emissions (e.g., by implementing emission trading systems), but dialogue about examining the various activities of transportation logistics between raw material mining and reaching the final customer, and eventually consumer, was rare. The same story applies to private car use: within the European Union and globally, emission cheating has been observed, along with massive corporate fraud, in diesel-powered cars, for example. As a result, fingers are now being pointed at the automotive industry (Oldenkamp et al. 2016; Li et al. 2018). However, the finger should also be pointed at politicians and citizens, since they continue to escape from reality with higher and more demanding vehicle emission standards, with the belief that further regulation will solve the outstanding problems (also partially concluded in Li et al. 2018). *"Technology development will solve these issues,"* many were thinking. However, the problems were actually solved by vehicle manufacturers taking shortcuts to success (where there are now a number of legal processes going on), which enabled much lower CO_2 and nitrogen emissions to be reported than what the reality was (Hotten 2015; Oldenkamp et al. 2016). Moreover, everyone believed that emissions were under control, even if diesel cars consumed nearly the same amount of diesel per distance driven as earlier models did (in real life). Air quality in European cities and roadsides also did not improve as much as expected (e.g., in terms of nitrogen oxides) by the implementation of tighter standards (Henschel et al. 2015). Should the behavioral and car usage systems have changed instead? Strategy drives structure, which in turn, gives the performance (results). Regulation itself is not the strategy, it is just legislation, which will be adapted to human and organizational behavior. For a long time, it has been known that diesel-powered car and truck engines are great problem in Europe, and should have a higher tax treatment than other alternatives (Mayeres and Proost 2001). In addition, studies exist from diesel car ban locations (like El-Zein et al. 2007), where the health of inhabitants has improved since the ban took place (the long-term effects were not that clear El-Zein et al. 2007).

One way of observing the argued situation in retrospective (transportation logistics environmental demands) is to analyze greenhouse gas (GHG) development in 28 countries in the European Union through official statistics (Fig. 1.1). It is true that from the base year 1990 (a time when still many countries were hugely polluting due to the

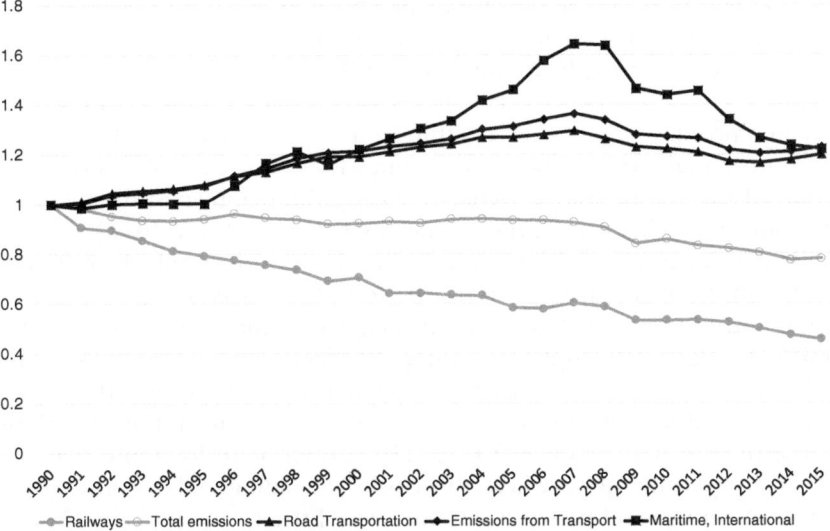

Fig. 1.1 Annual greenhouse gas (GHG) emissions within 28 countries of the European Union during time period 1990–2015 (indexed from million tons CO_2 equivalent, where $1990 = 1.00$) (*Source* (*data*) European Commission 2018a)

Soviet legacy), GHG pollutions overall have developed favorably—in 2015, they were 22.1% below the base year. However, this is the trend in overall activity. Transportation is very different compared with the general trend, and in the last observation year (2015), its share of the total GHGs was 26.6%. Since 1990, transportation has grown in GHG emissions by 23%. The highest polluter in this sub-sector is road transportation, which in the last observation year (2015) represented 72.1% of all transportation emissions. Road transportation GHG emissions have also grown by 20% over the years, despite all activity, technology development, implementations and legislation process accomplished. International maritime transport produces 11.4% of GHG emissions from the transportation sector, and it has also grown by 22.2% during the period (1990–2015) in its emissions. The only positive segment of the transportation sector are the railways, which have been able to accomplish a 54.4% decrease in emissions. This is of course partially because of technological progress (i.e., more energy efficient traction and trains overall) and implementations (such as higher use of electricity as traction), but also partially because of declining raw material transportation (especially coal) from the early 1990s and the lost market share in freight and passengers (especially in former Eastern Europe). In real life, some companies have started to utilize railways at a larger scale in order to combat high transportation CO_2 emissions. For example, BMW has reported that nearly 60% of vehicles leaving the manufacturing plant are transported for first part of the journey by railways (BMW 2017).

Of course, some might argue that the situation is good, since overall GHG emissions in Europe are on the long-term decline. As a counterargument, it could be stated that overall, across the entire world, GHG emissions are still strongly increasing. Readers could go and check the oldest measurement point, that of Mauna Loa (Hawaii) and the development of CO_2 emissions in this Pacific observatory from the late 1950s to the present date (NOAA 2018). The growth is really impressive and leaves no question as to where the world is heading. All actions, agreements, meetings and panels have done very little to the global CO_2 levels: it is still clearly increasing. Perhaps the achievement has been that it has not increased at a much higher rate.

A similar kind of process as with the regulation of 'private diesel cars' is now underway in freight transportation within the European Union. Initially, countries opened a 'wish box' of great hopes of further integration and globalization, only to discover that transportation activity had increased considerably (due to the much higher rate of trade and the low inventory, as well as high order frequency systems such as just in time). Now the situation is such that, in Europe, different countries have road use payments for trucks, and taxation is increasing for diesel fuels at gas stations almost annually. In addition, there is already agreement regarding incorporating transportation in CO_2 emission demands for the year 2030. These CO_2 reductions are massive to Finland, Sweden, Denmark and Germany, for example (around 40% reduction needed in 2030 from the 2005 level; European Commission 2018b). The maritime sector is not alone in here—in the year 2015 within Europe, stringent sulphur regulation was implemented in emission control areas (Baltic and North Sea), where sulphur content of used diesel was set at a maximum of 0.1% (IMO 2018a; Hilmola 2015). In 2015, a low sulphur level was also required for coastal areas of North America (together with some parts of US Caribbean Sea; IMO 2014). Sulphur content was globally decreased from 4.5% to 3.5% (in the year 2012).

In the same vein, China also implemented its own domestic sulphur emission control areas for three major sea port regions with cap of 0.5% (Liu et al. 2018); these started to be effective in some areas from the early part of 2016 and have enlarged in areal size in the years 2017 and 2018 (Hong 2017). Shipping is not like road transport, and its main problem is not CO_2 emission levels (3% from global CO_2 emissions; Stevens et al. 2015), since this is already very low compared with its absolute dominance in transported volumes (80% global modal share based on United Nations 2017). However, for a long time, shipping used inexpensive heavy fuel, which produced significant amounts of sulphur and nitrogen (Stevens et al. 2015; Lindstad et al. 2015). These in turn were, and still are, a very serious issue in populated areas due to their associated harmful health effects and, eventually, casualties. In North American coastal areas, implementation of demanding sulphur limits during the years 2012 and 2015 have also considerably decreased particular matter emissions, which in turn will mean improved health among the population living in close proximity to the coast (Kotchenruther 2017). Although, shipping-related CO_2 emissions are really low as compared with its transportation sector volume dominance at the global scale, IMO (2018b) have initially agreed that shipping CO_2 emissions ought to be reduced by at least 50% before the year 2050. As comparison year in emissions within here is 2008 level, and CO_2 emissions should reach their global peak as soon as possible.

Now, as decided by the International Maritime Organization (IMO), the sulphur cap for the entire world is going to be 0.5% from the year 2020 onwards. This will change maritime-based supply chains in the coming years and decades. The change is such that decision-makers need to think everything through in terms of costs and profitability, instead of scale and market share. Previously in shipping, fuel costs were not particularly high, when oil was trading 10–20 USD per barrel (in 1980s and 1990s); however, as prices have increased to the level of 50–100 USD per barrel and as legislation is forcing the use of lower-emitting grades, it will mean only one thing—the increase of variable costs. Previously, as the cost structure of maritime supply chains was capital intensive and the most worrying issues besides this were labour cost, profit taxation and payments for sea ports (as per visiting these to unload and/or load cargo). Now energy will enter the spotlight. Demands for shipping will not end here—CO_2 emissions will soon become part of the agenda, as has already happened with nitrogen emissions.

Of course, it is so that all forthcoming problems could be solved with money, in other words by large-scale investment in technology and new equipment. For example, sulphur could be washed with sea water by using ship-installed scrubbers (Ma et al. 2012; Yang et al. 2012; Lindstad et al. 2017). These scrubbers could be retrofitted to old ships as well. The only downside of this is that they would cost millions of euros, and their operation would not be free of charge (there is an ongoing demand for electricity and sulphur waste treatment at sea ports; see up-to-date estimates, Olaniyi et al. 2018). However, the use of scrubbers could buy time in order to use old fleet and heavy diesel oil. As Fig. 1.2 illustrates, the sulphur directive in 2015 (as well as in 2020) will either lead to higher capital investments or oil (fuel) costs. Reduced levels of sulphur in the environment will also mean higher overhead costs for some period of time as preparing for these changes. These changes will also have their effects on freight volume (most probably declining overall) and price (increasing). Similarly to sulphur regulations, the speed of vessels is having its effects on the maritime supply chain: higher ship speed and handling at sea ports will lead to better capital and labour turnaround (against revenue), but also, on the negative side, to a higher consumption of oil. Speed will also result in a higher number of port visits, and related payments. If oil is cheap

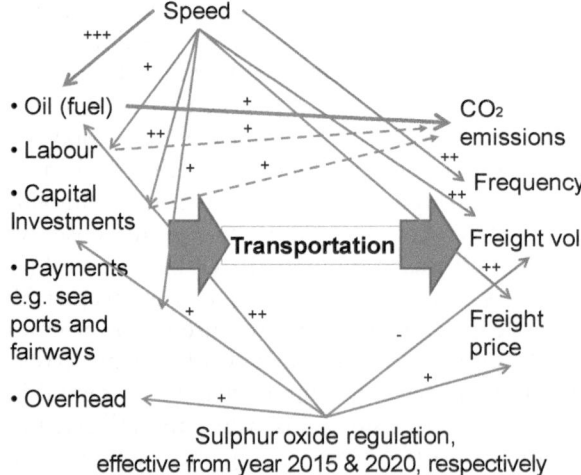

Fig. 1.2 Effects of speed and sulphur directive on inputs and outputs of maritime supply chain

and there are no environmental regulations concerning CO_2 emissions, it is really worthwhile for the shipping community to have higher speeds and a high consumption of heavy diesel. In some cases, routes were able to offer competitive prices, short lead times and high frequencies—and all this with a high freight volume gained. However, the logic of speed at shipping has lost its momentum, mostly due to high oil (fuel) costs and increasing environmental demands. In addition, globally, shipping has suffered from over-capacity in recent times, and this has resulted in the need to proceed slowly at deep sea in order to save costs and utilize the extra capacity for something. It will not be that surprising for the reader to discover in the following chapters that very short-distance sea journeys have won volumes at the Baltic Sea during the implementation of sulphur regulation. Companies are simply increasingly using the truck with semi-trailer combination to quickly load and unload cargo to ships (as forecasted in Notteboom 2011; Hilmola 2014, 2015), which connect to short-distance sea ports. This will of course favour speed, but increases the labour used and adds extra weight to ships. These factors will increase CO_2, even if the speed of operations improves.

It is rather difficult to understand the reason why trucks with semi-trailers were taking the market share in sulphur or environmental legislation tightening in Northern Europe. Figure 1.3 illustrates reasons for this further. Even if a container ship is polluting significantly less CO_2 and could be considered to be very environmentally friendly and sustainable due to high utilization ratios and slow speeds, it is not the winning

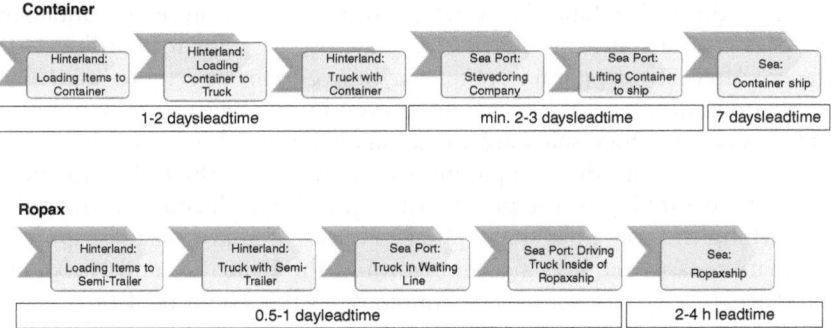

Fig. 1.3 Two maritime and unitized supply chains using either container ship or RoPax ship (from Northern Europe to Central Europe)

solution (Hilmola et al. 2015; Hilmola 2017). The reason is the container itself—even if it solves the handling issues of small-parcels/varying sized parcels, and eventually inefficiency at sea ports, it is still a troublesome concept for fast cycle-time supply chains (like those operating inside of Europe). First of all, containers (empty ones) need to be called to terminals or factories, where items are going to be loaded. Sometimes it is the case that empty containers travel hundred or even several hundreds of kilometers to reach the loading point. After this, the container is loaded full of items, either through automation or with a forklift (or even a more primitive device, such as a pump trolley). The more manual this operation is, the longer it takes (and the lower the quality). Thereafter, the container needs to be loaded on the truck trailer platform. It can be lifted, or alternatively the truck drives below loading platform in the terminal, and just takes the container on the trailer (It can be lifted, or alternatively the truck drives below container in the terminal, and just takes the container on the trailer (e.g. using legs besides container to have it lifted up in terminal). Thereafter, the road transportation combination starts its journey towards the sea port. The travel time is of course dependent on distance, the necessary breaks (based on legislation, personal and/or technical needs), traffic on the roads (and possible jams) as well as weather. It will take a minimum of one to two days that items reach sea port from the hinterland terminal (incl. loading items to container as shown in Fig. 1.3). At the sea port, the container is unloaded from the truck, and is put in the waiting area, where it is placed due to operations planning and customs issues (typically container supply chains at sea ports are required to enter custom zones since containers could travel to the EU, China, USA etc.). From this waiting area, containers are possibly moved using reach stackers close to ships for loading with cranes. Alternatively, gantry cranes may move them directly to ship. This whole operation takes a lot of time as even in Northern Europe volumes and feeder container ships are much smaller than those used by continental routes. Consider completing this activity for 400–600 containers. From the beginning of the process of terminal item loading to the point of reaching the container ship will take at least 3–5 days. Sea transport to destination easily requires a combination of the Baltic Sea and North Sea for at least 7 days (in some very limited cases this could be 4 or 5 days). Then containers need to go through foreign sea port formalities again, and a few days later (in the best-case scenario) they are free to be transported by hinterland to somewhere. Therefore, even to Central Europe, container shipping is not greatly used. Even if it were cost competitive, it would easily require two weeks' time, and for all this time all the items are held in inventory, which increases inventory holding costs and causes possible quality damage. In addition, customers are required to wait longer, and, in rapidly changing markets, items lose their value. This system is also very dependent on the availability of different machines of container handling, shipping schedules (low frequencies, typically weekly or bi-weekly) and labour with various union contracts as well as skillsets.

Some readers might wonder, why these maritime supply chain examples do not contain the railway. It is as environmentally friendly (or in some situations, even better) as container shipping, and favoured in public appearances of politicians everywhere. The reason is very simple: it is even slower than the truck used to transport containers to the container ship. The container train contains 20–40 wagons (this differs between European countries, and also as a result of transportation demand), and in each one of those, there are possibly two containers. It varies as to whether trains can leave directly from factories (those with sufficient volume) or alternatively container shall start first with a truck, and continue to nearby railway terminal. Railway transportation is by itself complex combination of different issues, where traction engine(s) is needed, free capacity is needed to be booked from railway network, container wagons are needed and then lifting at terminal area is required and so on. At the sea port, the situation is similar; a railway terminal is needed and then the operations with regard to the container ship are as earlier described. Railways and container ships are really the solution in the environmental sense, but together they are far too slow. In Fig. 1.3, it shows that operations prior to the sea port took in the truck case 1–2 days. With railways, this time is at least 3–4 days (of course, there might be situation where this lead time is close to trucks, but it requires a high volume railway route with numerous daily departures). Lower emission maritime supply chains exist, but they will require items moving much more slowly. Railways are typically used if the distance from sea port to the hinterland point is between 500 and 1000 km (at least). Shorter distances could be utilized, but the only successful examples are from Sweden (so called dry port concept; Roso et al. 2009; Roso and Andersson 2017), and the dry port system in Sweden has the challenge of transporting semi-trailers over a large scale (Woxenius and Bergqvist 2011; Bergqvist and Cullinane 2017) since it is merely a container-based system.

The alternative for this, which is also truck based, is to load items at the terminal to a semi-trailer (Fig. 1.3, lower part), which comes together with the truck (or then empty semi-trailer is taken from somewhere, and it is picked by truck). This semi-trailer combination will proceed through the hinterland to a sea port of a RoPax ship (or RoRo), where it enters the waiting line. Typically, a couple of hours at the waiting line is sufficient that the truck driver is able to drive truck and semi-trailer combination inside of ship. Loading operations at RoPax ships are really straight forward, and do not take a long time. As an example, in Finland, RoPax ships will in most cases enter EU area countries, and then the custom formalities are very simple (or indeed, they do not exist at all). RoPax ships typically operate in fast mode at the sea, they consume a lot of diesel (fuel) and they pollute a lot. However, they benefit cargo a lot too. The service is fast, and it is not greatly dependent on sea port labour (very little indeed), and various union issues can be avoided (e.g., stevedoring strikes). In the best situation of under a day, cargo is for example from Finnish manufacturing unit or logistics terminal within the road network of Estonia. Within day or two out of this, most of the destinations in Central Europe can be reached (using Baltic roads). This system of course produces a high level of pollution and could be too expensive, but in reality, it is not. Trucking is always under a vast amount of competition (due to deregulation processes taking place in the European Union), so there are always options available, and prices are reasonable. Drivers come from a diverse set of countries, where salaries are competitive. In addition, diesel prices differ inside of European Union countries considerably (reader may check diesel prices in Poland, three Baltic States, Finland and Sweden e.g. in DKV 2018; differences are still there). Even if environmental demands substantially hit the maritime sector, it will not have a significant effect on very short-sea shipping routes (max. 100–150 km). Hinterland taxes for road transport would need to greatly increase in order for this system not to be competitive. It will also provide many other positive business impacts: inventories are really low and tied capital is much lower, the quality of operations is higher (punctuality and lower product damage), RoPax ships typically have numerous daily departures, and it enables a door-to-door chain with few controllable actors. As Woxenius and Bergqvist (2011) concluded from research into the Swedish situation, containers deal in long-distance continental transports and the time-frame is in weeks, whereas trucks with semi-trailers are operating within a time frame of hours. Precision in transport service is also very different since in containers it is measured in days, whilst in semi-trailers it is hours (Woxenius and Bergqvist 2011).

The following chapters will review the changes that took place due to sulphur regulation implementation at Baltic Sea. This is examined mostly through combining, compiling and analysing second-hand data from different sources. Most datasets already have three years of data since the change took place, so interpretation for the medium term can be made. Chapter 2 starts this analysis by examining three countries of interest at the northern Baltic Sea: Estonia, Finland and Sweden. The macro-economy and trade situation is analysed, along with the oil price dynamics that have taken place in the previous two decades. Significant oil price decline in the latter part of 2014 and in 2015 considerably softened the effects of sulphur regulation in the Baltic Sea, but did not remove them completely (Zis and Psaraftis 2017; Hilmola et al. 2017). It could be considered as a short-term financial aid. Chapter 3, in turn, describes what happened to the unitized cargo flows of Finland to three main destinations: Estonia, Germany and Sweden. It is clearly shown that the Estonian route benefitted in volume terms from the implementation of sulphur regulations, while the German and Swedish routes were hurt. The situation of cargo handling, and particularly unitized cargo handling, at the sea ports of these three countries (Estonia, Finland and Sweden) is dealt with in Chapter 4. As an interpretation of results, it could be stated that trucking, and particularly using trucks accompanied with semi-trailers, has increased significantly over the years as these are loaded onto short-sea shipping routes. This is not only the situation in Finland and Estonia, but also in Sweden. In other respects, sea ports have experienced challenges regarding volume development (tons) overall, but also have clear weaknesses in container handling. However, maritime supply chains using trucks and semi-trailers have clearly been the benefitting party.

Company-specific issues of mostly Finland and Estonia are dealt with in Chapter 5. It could be identified that shipping companies are not growing that much, and actually, revenues are experiencing a small, but long-term decline. In addition, asset values are declining. It seems that shipping companies, which invested mostly in scrubbers to tackle sulphur regulation, produced best the financial results during the years 2015–2017. However, opposite was also the case, as shipping company in bulk segment performed pretty well in the period, even if mostly relying on low sulphur content diesel. Two publicly traded hinterland transports companies were also analysed (due to clear change in the mindset and use of new routes), and in here, it seems that the trucking companies were able to show a continuum in revenue growth. However, the company's profitability did not change much over the years. Chapter 5 also analyses the situation of using

Liquefied Natural Gas (LNG) from the sea port perspective (LNG terminal, which has some years of operation), and also by analysing LNG price development in different continents. This is owing to the fact that many analyzed shipping companies have invested in LNG ships or are in the process of building them through shipyards; in addition, private sector interviews revealed interest for this fuel and technology as part of tackling change. LNG is a promising new source of energy to be used in maritime supply chains, but it has its technical, infrastructure, emission and business risks. Multiyear surveys completed primarily by Finnish and Swedish companies are analyzed in Chapter 6, and the last survey of 2015 includes also Estonian responses. An introduction and analysis of three simulation models follows in Chapter 7. The book ends with concluding discussion in Chapter 8.

Critical questions for readers (book will facilitate and/or provide answers):

- Are you aware of the cost effects of the coming sulphur regulations on supply chains?
- How much is your company is dependent on maritime supply chains?
- What is the form and performance of maritime supply chain, if the maritime component is being minimized? What is the role of different hinterland transport modes?
- What is your strategy to tackle the coming sulphur regulation change?
- What is the CO_2 emission level (direct result of diesel consumption) of your maritime supply chain?
- How will different routes (e.g., export, import, semi-finished goods or raw materials) experience the coming regulation cost increases? Can use of more cost-efficient routes be increased in the short amount of time available?
- Is your supply chain network more regional rather than concentrating globally on some products in dedicated locations? If the network is globally focused, are you aware of transportation cost implications?
- What kind of environmental investments could make such that your organization is better prepared and more competitive in the new low-sulphur world?

References

Bergqvist, R., & Cullinane, K. (2017). Port privatization in Sweden: Domestic realism in the face of global hype. *Research in Transportation Business and Management, 22,* 224–231.

BMW. (2017). *Sustainable value report 2016.* Munich, Germany. Available at https://www.bmwgroup.com/content/dam/bmw-group-websites/bmwgroup_com/ir/downloads/en/2016/BMW-Group-SustainableValueReport-2016–EN.pdf. Retrieved 27 November 2017.

Bowman, R. (2014, June 3). Third-party logistics providers are shrinking in number, growing in size. *Forbes.* Available at https://www.forbes.com/sites/robertbowman/2014/06/03/third-party-logistics-providers-are-shrinking-in-number-growing-in-size/#7eb83252426e. Retrieved 29 March 2018.

DKV. (2018). *Interactive diesel price map.* Available at https://www.dkv-euros-ervice.com/DKVMaps/. Retrieved 29 March 2018.

El-Zein, A., Nuwayhid, I., El-Fadel, M., & Mroueh, S. (2007). Did a ban on diesel-fuel reduce emergency respiratory admissions for children? *Science of the Total Environment, 384,* 134–140.

European Commission. (2018a). *Statistical pocketbook 2017, mobility and transport.* Brussels: European Commission. Available at https://ec.europa.eu/transport/facts-fundings/statistics/pocketbook-2017_en. Retrieved 22 March 2018.

European Commission. (2018b). *Proposal for an effort sharing regulation 2021–2030.* EU Action. Available at https://ec.europa.eu/clima/policies/effort/proposal_en. Retrieved 29 March 2018.

Henschel, S., Tertre, A. L., Atkinson, R. W., Querol, X., Pandolfi, M., Zeka, A., et al. (2015). Trends of nitrogen oxides in ambient air in nine European cities between 1999 and 2010. *Atmospheric Environment, 117,* 234–241.

Hilmola, O.-P. (2014). Growth drivers of Finnish-Estonian general cargo transports. *Fennia—International Journal of Geography, 192*(2), 100–119.

Hilmola, O.-P. (2015). Shipping sulphur regulation, freight transportation prices and diesel markets in the Baltic Sea region. *International Journal of Energy Sector Management, 9*(1), 120–132.

Hilmola, O.-P. (2017). Transport modes and intermodality. In J. Carlton, P. Jukes, & Y.-S. Choo (Eds.), *Encyclopedia of marine and offshore engineering.* New York: Chapter for the Wiley Publishers.

Hilmola, O.-P., Kiisler, A., & Hilletofth, P. (2017). Cabotage and sulphur regulation change: Cost effects to Northern Europe. *International Journal of Business and Systems Research, 11*(4), 417–428.

Hilmola, O.-P., Lorentz, H., & Rhoades, D. L. (2015). New environmental demands and the future of Helsinki-Tallinn freight route. *Maritime Economics and Logistics, 17*(2), 198–220.

Hong, Y. (2017). Maritime transport and sulphur regulation: Background and remedies [航运"限硫令"出台背景及应对措施]. *Marine Equipment Materials & Marketing* [船舶物资与市场]. Issue 4, pp. 46–49.

Hotten, R. (2015, December 10). Volkswagen: The scandal explained. *BBC News.* Available at http://www.bbc.com/news/business-34324772. Retrieved 20 March 2018.

IMO. (2014). *Frequently asked questions: Sulphur limits in emission control areas from 1 January 2015.* International Maritime Organization. Available at http://www.imo.org/en/MediaCentre/HotTopics/GHG/Documents/sulphur%20limits%20FAQ.pdf. Retrieved 26 April 2018.

IMO. (2018a). *Sulphur oxides (SOx) and particulate matter (PM)—Regulation 14.* Available at http://www.imo.org/en/OurWork/Environment/PollutionPrevention/AirPollution/Pages/Sulphur-oxides-(SOx)-%E2%80%93-Regulation-14.aspx. Retrieved 19 February 2018.

IMO. (2018b). *UN body adopts climate change strategy for shipping.* Available at http://www.imo.org/en/MediaCentre/PressBriefings/Pages/06GHGinitialstrategy.aspx. Retrieved 27 June 2018.

Kotchenruther, R. A. (2017). The effects of marine vessel fuel sulfur regulations on ambient $PM_{2.5}$ at coastal and near coastal monitoring sites in the U.S. *Atmospheric Environment, 151,* 52–61.

Li, L., McMurray, A., Xue, J., Liu, Z., & Sy, M. (2018). Industry-wide corporate fraud: The truth behind the Volkswagen scandal. *Journal of Cleaner Production, 172,* 3167–3175.

Lindstad, H., Eskeland, G. S., Psaraftis, H., Sandaas, I., & Strømman, A. H. (2015). Maritime shipping and emissions: A three-layered, damage-based approach. *Ocean Engineering, 110,* 94–101.

Lindstad, H. E., Rehn, C. F., & Eskeland, G. S. (2017). Sulphur abatement globally in maritime shipping. *Transportation Research Part D, 57,* 303–313.

Liu, H., Meng, Z.-H., Shang, Y., Lv, Z.-F., Jin, X.-X., Fu, M.-L., et al. (2018). Shipping emission forecasts and cost-benefit analysis of China ports and key regions' control. *Environmental Pollution, 236,* 49–59.

Ma, H., Koen, S., Xavier, R. P., & Nigel, T. (2012). Well-to-wake energy and greenhouse gas analysis of Sox abatement options for the marine industry. *Transportation Research Part D, 17*(7), 301–308.

Mayeres, I., & Proost, S. (2001). Should diesel cars in Europe be discouraged? *Regional Science and Urban Economics, 31,* 453–470.

NOAA. (2018). *Atmospheric CO_2 at Mauna Loa observatory, full record.* U.S. Department of Commerce, National Oceanic & Atmospheric Administration. Earth System Research Laboratory. Available at https://www.esrl.noaa.gov/gmd/ccgg/trends/full.html. Retrieved 22 March 2018.

Notteboom, T. (2011, April). The impact of low sulphur fuel requirements in shipping on the competitiveness of roro shipping in Northern Europe. *WMU Journal of Maritime Affairs, 10*(1), 63–95.

Olaniyi, E. O., Atari, S., & Prause, G. (2018). Maritime energy contracting for clean shipping. *Transport and Telecommunication, 19*(1), 31–44.

Oldenkamp, R., Zelm, R., & Huijbregts, M. A. J. (2016). Valuing the human health damage caused by the fraud of Volkswagen. *Environmental Pollution, 212,* 121–127.

Roso, V., & Andersson, D. (2017). Dry ports and logistics platforms. In J. Carlton, P. Jukes, & Y.-S. Choo (Eds.), *Encyclopedia of marine and off-shore engineering*. New York: Chapter for the Wiley Publishers. https://doi.org/10.1002/9781118476406.emoe201.

Roso, V., Woxenius, J., & Lumsden, K. (2009). The dry port concept: Connecting container seaports with the hinterland. *Journal of Transport Geography, 17,* 338–345.

Stevens, L., Sys, C., Valenslander, T., & van Hassel, E. (2015). Is new emission legislation stimulating the implementations of sustainable and energy-efficient maritime technologies? *Research in Transportation Business and Management, 17,* 14–25.

United Nations. (2017). *Review of maritime transport*. Geneva: Unctad, United Nations.

Woxenius, J., & Bergqvist, R. (2011). Comparing maritime containers and semi-trailers in the context of hinterland transport by rail. *Journal of Transport Geography, 19,* 680–688.

Yang, Z. I., Zhang, D., Caglayan, O., Jenkinson, I. D., Bonsall, S., Wang, J., et al. (2012). Selection of technologies for reducing shipping Nox and Sox emissions. *Transportation Research Part D, 17*(7), 478–486.

Zis, T., & Psaraftis, H. N. (2017). The implications of the new sulphur limits on the European Ro-Ro sector. *Transportation Research Part D, 52,* 185–201.

CHAPTER 2

General Economic and Trade Environment

Abstract It is necessary to review the operating and economic environment of Northern Europe in order to understand the implementation environment of the very low sulphur level. The economies of Estonia, Finland and Sweden are in general wealthy, and the two of them have shown clear medium-term growth in GDP. During the examination period, all three countries became net importers, even if the earlier situation was rather different, with constant trade surpluses (in Finland and Sweden). In overall trade, only Finland is showing some long-term weakness, whilst Sweden and Estonia are still experiencing steady growth. When tight restrictions on sulphur levels were implemented in 2015, oil prices were significantly declining, and this helped the region to avoid big cost increases. However, in 2017, oil prices started to show an upwards movement. It is clear that lower sulphur maritime diesel oil has a far higher price tag, and in 2017, its price increased at a higher rate than conventional maritime diesel oil.

Keywords GDP · Trade · Estonia · Finland · Sweden · Oil · Diesel

2.1 INTRODUCTION

In year 2015 sulphur regulation change countries that were carrying most of the liability and costs, were small (these could be considered together as a testing laboratory for forthcoming global implementation).

© The Author(s) 2019   17
O.-P. Hilmola, *The Sulphur Cap in Maritime Supply Chains*,
https://doi.org/10.1007/978-3-319-98545-9_2

Consider the three countries analysed in this chapter and dealt with in this book: Sweden has a population of around 10 million inhabitants, Finland in turn 5.5 million and Estonia 1.3 million. Other countries the northern side of the Baltic Sea are also rather small, such as Latvia and Lithuania. Only Poland and Germany are bigger, but regulation change effects on their economies were not that significant since short-distance hinterland access exists to key markets, and the Mediterranean region could always be used as another access point. Of course, sulphur regulation at the Baltic Sea also concerned Russia through its important St. Petersburg sea port complex, but this country also has numerous other locations, which it can use as an access points for imports and exports. The Black Sea is one such access point. In addition, hinterland and pipeline transport alternatives exist to both Central Europe and Asia.

What makes this book and the evaluation of these three small countries fruitful, is their imminent need to have functioning and cost-efficient logistics connections. From these small countries, Estonia is one of the most international trade-intensive countries in Europe and the world. Based on OECD (2016), Estonia traded goods and services worth of 164.4% of GDP (imports having a somewhat higher role since it is also in physical goods trade). From Swedish GDP, the international trade of goods and services in the year 2014 accounted for 85.3%, and from Finnish GDP they accounted in turn 76.6%. On average in European Union during 2014, this measure was slightly above the performance of Finland. In all OECD countries on the average shares of trade are lower, somewhere below 60%. So, all of these three countries are good cases for analysis due to their sensitivity concerning trade. Changes in trade in these countries should be well present as thinking about implemented demanding environmental regulation. All of these three countries are entirely dependent on international trade and smooth-running and cost-efficient logistics. Maritime supply chains also play an important role in all of these countries.

This chapter is structured as follows: in Sect. 2.2, GDP development of the three countries in question is analyzed. Thereafter, in Sect. 2.3, trade development, and the situation of imports and exports is dealt with. Oil price development in long-term, and within both USD and euro currencies is introduced in Sect. 2.4. Since environmental demands are having impacts on the refinery process of fuels, the cost difference and latest development concerning different sulphur grades is shown in Sect. 2.5.

2.2   DEVELOPMENT OF GDP IN ESTONIA, FINLAND AND SWEDEN

In terms of economic development, the three examined countries have their differences. Sweden has for a long time been one of the wealthiest countries in the world (56,300 USD per capita GDP in year 2016) and is exemplary in economics in many respects. Although Sweden already had a high GDP in early 2000, it has been able to nearly double nominal GDP by the end of the 2017 (see Fig. 2.1). Finland, in turn, is like a little brother to Sweden (45,700 USD per capita GDP in 2016) and has always followed Sweden in economic terms. However, in early 2000 when the growth period started, these two countries were hovering at same economic prosperity level, and Finland grew similarly with Sweden until the credit crunch of 2009. Thereafter, even in nominal terms it has been very difficult for the Finnish economy to grow. The reasons are numerous; however, losing competitiveness in export industries and manufacturing as well as the effects of an ageing society effects are

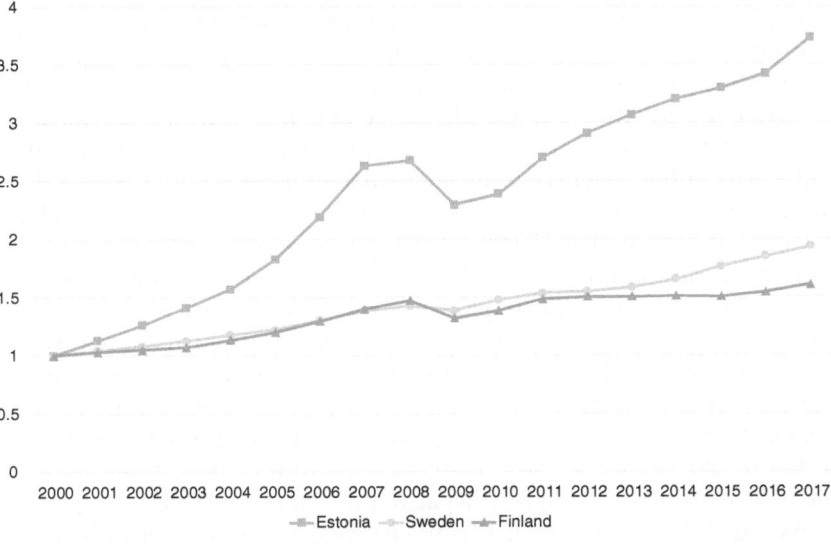

Fig. 2.1 Development of nominal GDPs in three countries during the period of 2000–2017 (indexed, where 2000 = 1.000) (*Source (data)* Statistics Estonia 2018; Statistics Finland 2018; Statistics Sweden 2018)

the typical points raised. At the end of 2017, the Finnish economy was around 1.61 times higher compared with what it was in the base period (Fig. 2.1).

From these three countries in question, Estonia (17,800 USD per capita GDP in 2016) has been exemplary in its economic reforms in the transition process from a Soviet-ruled state to fully sovereign country. Its taxation (e.g., flat-rate in income tax, no taxes from companies, if dividends are not shared to shareholders), free market approach and lower social security net have provided a platform for the economy to grow. Only in very recent years has Estonia started to widen its social benefits for different inhabitant groups (such as families), and the country is somehow gradually absorbing the economic model of Finland and Sweden. Anyway, in nominal terms, the Estonian economy has grown enormously (inflation has been present too), and from the year 2000 its economy, in nominal terms, was 3.73 times larger in 2017. Even if the growth rate is very high, it should be remembered that Estonian economy overall is small (in absolute terms). It is, in overall size, ten times smaller than the Finnish economy. The Swedish economy in turn is twice as big as the Finnish economy. This difference could be in part explained by population differences, but also with actual wealth effects.

From these countries, two are using euros as their currency (Finland and Estonia), but Sweden still uses its own krona. Apart from joining the eurozone in early 2011, Estonia has been able to show growth; however, this is not the case for Finland. It is a question of competition, and as Finland has a higher salary level, and other salary-related costs, it is difficult to compete with those in a strong currency environment. Therefore, the euro has created challenges for Finnish economy. In recent years, Finnish salaries have been cut (2016–2017) for the first time in decades due to difficulties of finding growth for export industries, and also as a result of increasing governmental budget deficits due to public sector wages. Sweden as the outsider of common European currency area, in turn, has been able to utilize its own currency as an economic asset. It has been flexible in times of lower growth (weakening just as needed). In addition, the Swedish Central Bank was one of the first Central Banks in the world to introduce slightly negative interest rates in 2015.

As Fig. 2.1 reports nominal GDP changes, it is important to highlight that real inflation-adjusted growth has been much slower. For example, Finland is still in real GDP terms below the level of 2008. Instead of growing in the entire observation period by more than 1.61 times, the real growth was around 1.25. A similar lower real GDP is also present in Swedish numbers; Fig. 2.1 reports that nominal GDP grew nearly by two times in the observation period, but real GDP grew by 1.45 times. However, the Swedish economy has been stronger in the period following 2009, as it has grown to 1.181 times higher compared with 2017. Estonia shows the best growth from these three countries, however, it suffers from inflation: In real terms, its GDP grew during years 2000–2017 by factor of 1.77. Growth following the 2009 recession has been fueled by higher inflation since a GDP growth rate of 1.39 times from 2008 is, in real terms, 1.12. So, actually, Sweden has been showing a stronger economy since the global credit crunch, which is rather surprising. This is made understandable by examining Fig. 2.1 further—as Estonia (−14.36%) and Finland (−10.46%) both recorded in nominal terms big drops in GDP s during 2009, Sweden was able to continue with manageable decline of 2.93%. The reason for this was it having its own currency, which was able to show flexibility as needed (Estonia had its own currency, but it was considered as strong one with small daily fluctuations). The same situation was also elsewhere in Europe, in countries where competition among was present and where the currency was flexible. A good example is Poland, which in 2009 showed growth of GDP in national currency terms. This is in line with Stiglitz (2018), who clearly illustrated that economic development during 2007–2015 in Non-eurozone Europe has been much better compared with currency union countries.

2.3   Development of Trade in Estonia, Finland and Sweden

Earlier, and prior to the global credit crunch of 2009, Sweden (Fig. 2.4) and Finland (Fig. 2.3) were both showing strong trade performance, where trade surpluses persisted, and were actually substantial. Even in this era, Estonian trade was showing deficits, and it has been in this mode in the entire observation period of Fig. 2.2. The size of Estonian economy is small, and it can tolerate trade deficits, which could be covered with services such as logistics and tourism. Of course, all deficit countries are dependent on external debt, but in the Estonian case it is very small (one of the lowest governmental debt to GDP performance ratios in all of Europe).

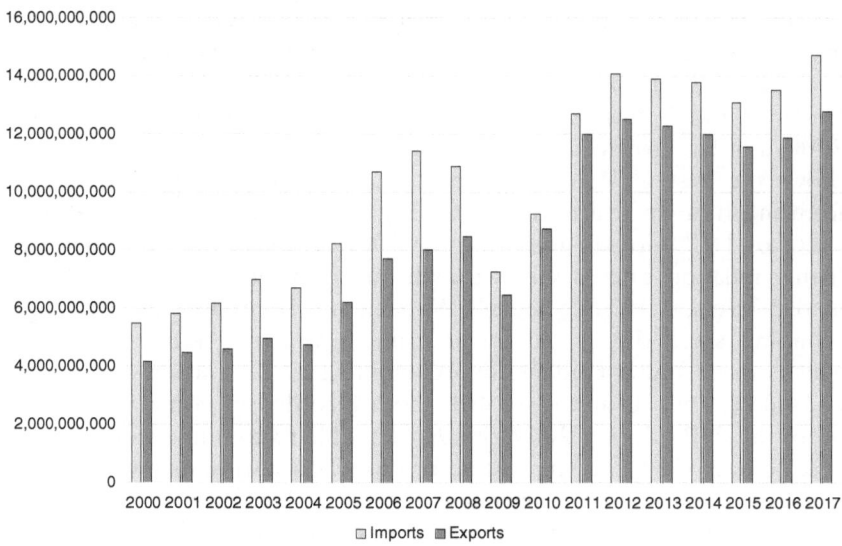

Fig. 2.2 Development of import and export of Estonia during period of 2000–2017 (in EUR) (*Source (data)* Statistics Estonia 2018)

What is interesting in trade development, is the gradual turning of all three countries into trade deficit nations. Finland turned to such after 2010, and Sweden has, in very recent years developed into one (with small deficit during both 2016 and 2017). Typically, it is very difficult to quickly turn deficit into surpluses as a turn to negative has taken place. In some countries, deficits do not matter that much, if capital is available, if the service economy covering deficits (banking, logistics, tourism, and/or retail) and/or if capital markets and real-estate are booming (creating wealth effects). This has been the case in Estonia and Sweden, and it could be identified from Figs. 2.2 and 2.4 that overall trade has grown pretty well, and imports have continued to increase. This is not the case in Finland, where the domestic market has been relatively weak during years 2009–2016 (since in 2017, there was clear turnaround in both exports and imports).

During the observation period of Fig. 2.2, Estonian total trade increased by 2.85 times compared with the base period. Even if imports were consistently higher than exports, it was exports that grew most in this period—around three times more. With the point of comparison set

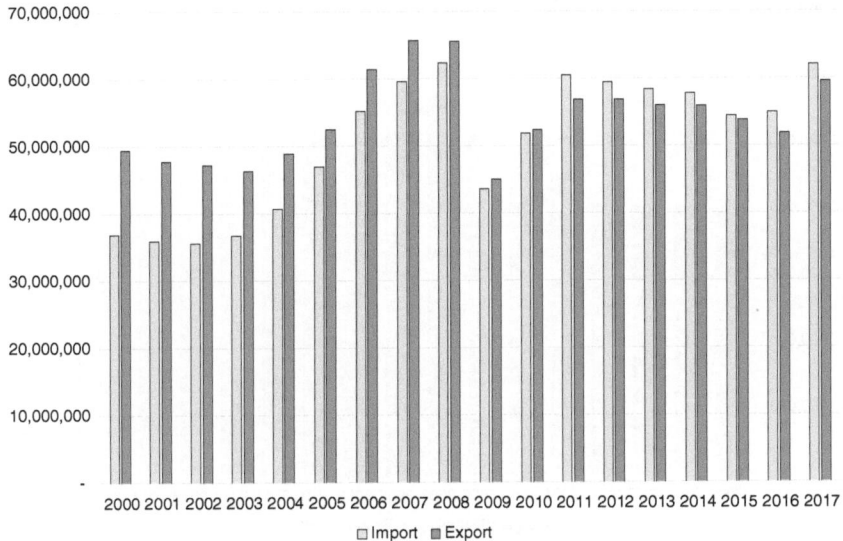

Fig. 2.3 Development of import and export of Finland during the period of 2000–2017 (in EUR) (*Source (data)* Finnish Customs 2018)

to the year 2008, total trade grew by 42.1% until 2017. Both imports and exports have grown, but again exports showed higher growth rate (50.9% vs. 35.2%).

Trade volume development in Finland is a different story from that of Estonia and Sweden. As concluded earlier, after the year 2009 growth has largely been lacking (apart from in 2017). As Fig. 2.3 shows, both import and export are, in euro terms, below the 2008 level, even in the last observation year 2017. Imports have slightly declined from 2008 (−0.56%), but exports in turn have quite substantially declined (−9.19%). Total trade is around 5% lower in 2017 compared with what it was in 2008. The situation is not so sad for exports, if volumes are observed (in tons/kg). In fact, export prices have substantially declined from 2008, but in turn, volumes have increased. Therefore, bulky trade logistics are still doing well (especially in exports). In weight terms, exports are somewhat above the 2008 level. However, it should be remembered that for the years 2015–2017, the Finnish government halved fairway due payments to directly aid shipping (in the face of sulphur regulation), and also removed track tax in order to facilitate railway usage

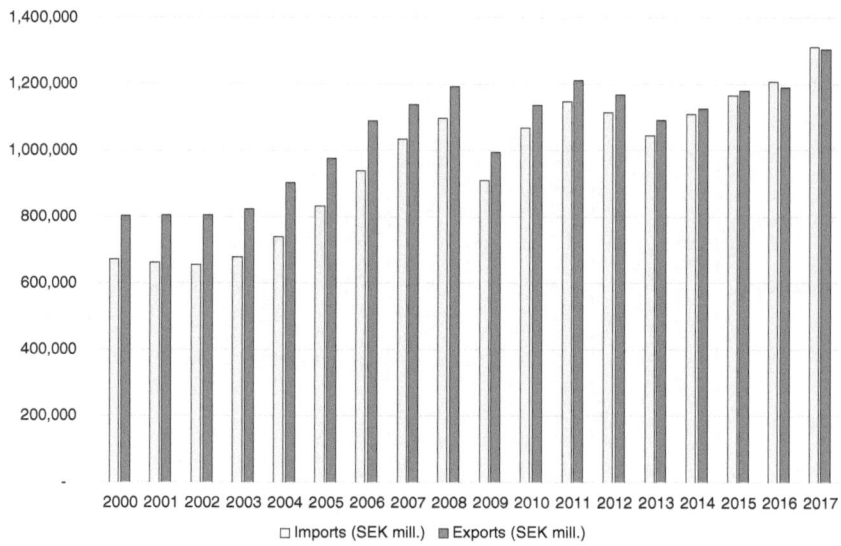

Fig. 2.4 Development of import and export of Sweden during period of 2000–2017 (in million SEK) (*Source (data)* Statistics Sweden 2018)

(contemporary measures continued in the year 2018)—these were evaluated to cost the government 55.7 million EUR per year (Ministry of Transport and Communications 2014).

Finland of course has shown growth in trade, but the observation period then needs to be enlarged to start from year 2000. In this situation, total trade has grown 40.9%, and imports have grown much more than exports (68.4% vs. 20.3%). Import prices have of course increased, and in weight terms, imports have actually declined over the years.

Swedish total trade has increased from 2000 by 77.4% (see Fig. 2.4), and imports have shown more growth than exports (95.3% vs. 62.5%). The situation has remained the same since 2008. Total trade has grown by 14.3%, and growth of imports is twice as high as in exports (19.6% vs. 9.4%). It is of course always a question as to how long imports can increase for. In the Swedish situation, it could take years to notice that this could be the possible macro-economic problem. The wealth effect in terms of real-estate and equity markets has been rather beneficial in Sweden, and this provides support for further domestic consumption. In addition, interest rates are very low in Sweden, and the requirements to pay back mortgages, for example, are really relaxed (with very long payback times).

Demanding sulphur regulations were implemented for the Baltic and North Sea at the beginning of the 2015. Since this change will almost automatically increase fuel costs, its effects on imports and exports as well as overall trade are worthy of observation. Of course, many issues affect trade, and logistics accessibility and costs are some of them, however, macro-economic issues are also a vital part of the equation. Upon analyzing the situation in these three countries, it could be concluded that in Estonia and Finland imports, exports and overall trade declined in 2015. From the level in 2014 level, it was down overall by 4.4% in Estonia and 4.7% in Finland. In both countries, imports fell by a higher rate in 2015. Decline continued in Finland during 2016, when overall trade level was 6% lower as compared with 2014 level. The situation in Estonia was a bit better, as after some recovery overall trade was 1.5% lower than in 2014. In Sweden, 2015 and 2016 did not bring any declines—trade actually grew rather well overall in 2015 (+4.9% compared with the year 2014), and growth continued in 2016 (but more slowly). As 2017 showed recovery all over Europe, all three analyzed countries demonstrated really nice growth rates for this year.

2.4   Increasing Oil Prices: Driver of Harm in Sulphur Regulation

Energy prices, and particularly the price of oil, have been on a roller-coaster ride over the previous decades (Fig. 2.5). However, what is typically forgotten is the clear trend experienced in prices—they move clearly upwards, but of course have volatile spikes both up and down. For example, in US dollar terms, in the 1990s oil was cheap; it even touched somewhere around 9 USD per barrel levels in late 1998 (European Brent quality). However, prices also spiked in the early 1990s with oil price was above 40 USD per barrel. The average price of oil in the 1990s was around 18.4 USD per barrel. In the following decade, these prices were a thing of the past. Oil nearly reached the 40 USD per barrel level once again in the year 2000, and by July 2008 this had increased to more than 140 USD per barrel. From the lows of 1998 to the all-time highs of 2008, oil had climbed by 15.8 times in US dollar terms (1481.9% increase). Even if oil prices have sustained their high level in

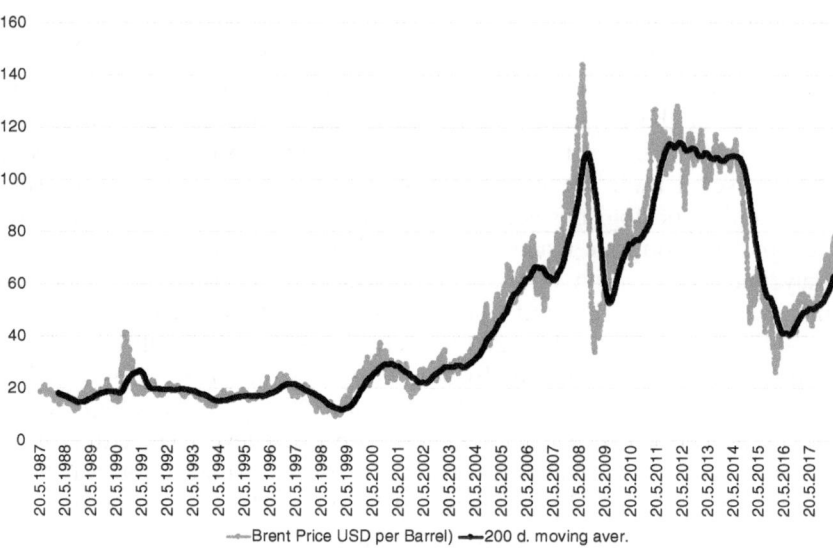

160

140

120

100

80

60

40

20

0

20.5.1987 20.5.1988 20.5.1989 20.5.1990 20.5.1991 20.5.1992 20.5.1993 20.5.1994 20.5.1995 20.5.1996 20.5.1997 20.5.1998 20.5.1999 20.5.2000 20.5.2001 20.5.2002 20.5.2003 20.5.2004 20.5.2005 20.5.2006 20.5.2007 20.5.2008 20.5.2009 20.5.2010 20.5.2011 20.5.2012 20.5.2013 20.5.2014 20.5.2015 20.5.2016 20.5.2017

Brent Price USD per Barrel) 200 d. moving aver.

Fig. 2.5 Oil price (Brent quality) in USD during the period of May 1987–May 2018 (*Source (data)* EIA 2018)

the following decade, the price trend has not shown such an aggressive increase compared with the decade before. During very recent years (since the Ukrainian dispute), starting from late 2014 to 2015, there has been an aggressive downwards movement. However, even this sudden decline did not lead to 20 USD per barrel oil for longer period of time. Oil prices have clearly recovered in 2017, and this has continued in the first half of 2018. Brent is again approaching 80 USD level. In mid-May 2018, the price of oil had increased from the beginning of the year by 17.3%.

Daily prices are one methods of observing oil expenses, but this is often too short-term a measure. For business using a lot of oil (or diesel products), a much better measure is a moving average of the previous 50/100/150/200 days. It tells the real price, since it is difficult—if not impossible—to catch lows in oil purchase decisions monthly or weekly. As Fig. 2.5 shows, the oil price peak in the year 2008 was very quickly formed and short-lived. Moving average of 200 days reached 110 USD in late 2008 (September), and lost its value rapidly as the global credit crunch sent shockwaves to economies around the world. After recovery and large-scale quantitative easing by central banks, the price of oil actually recovered its old high levels. This was not in daily price terms, but rather in moving average of 200 days. In fact, in early May 2012, new highs of 114.3 USD per barrel were recorded. Oil was actually big concern for logistics companies for the entire period of 2011–2014. This was of course greatly concerning in the Baltic Sea area since the strict sulphur level resctrictions of 2015 were on the horizon and oil prices were already high. If they were to have been sustained, then the price change to a very low-level ulphur diesel would have been very costly.

The implementation of a strict sulphur content in maritime supply chains turned out to happen at a fortuitous time in Northern Europe as oil prices were crashing simultaneously as more expensive diesel oil was required to be used in ships. This of course was a very short-lived period as it only lasted for two years (2015–2016). On an annual basis in USD, Brent was 47.1% cheaper in 2015 compared to 2014, and in 2016 it was 16.6% lower in priced compared with the previous year. In 2017, oil prices were getting clearly again, and direct diesel costs were getting higher. The annual average price in 2017 was 24% higher in USD terms than the year before. In the first four and half months of 2018, this price hike has continued. Of course, it would have been smart for companies to have invested during these cheap years without rushing to scrubber technology and using low-sulphur diesel oil from the very beginning. Companies would have simply received more reliable scrubber technology, since implementations and experience were increasing all the time. In retrospect, this would have been the perfect strategy to tackle this huge problem. Coming global year 2020 change could learn from this. Even if oil prices increase for while, and were at the level of 80–100 USD per barrel on the eve of the 2020 global implementation, it may still be the best option just to 'wait and see.' Cleaning technologies develop all the time, and devices such as scrubbers could experience scale effects. Therefore, rushing to install scrubbers would not necessarily be the best option.

From the European perspective, oil prices have been less volatile as the euro currency has shown better relative strength compared with the US dollar. Figure 2.6 illustrates Brent quality prices from early 1999 (as from this year onwards euro was introduced as bank calculated accounting currency, first monetary use implementation was made in the early 2002). In the beginning of 1999, one barrel cost 8.62 EUR. In July 2008, the price increased to 90.74 EUR per barrel. The change was of course 10.5 times higher, but it was not as high as the cost in USD. The highest price in the observation period in Fig. 2.6 was reached in

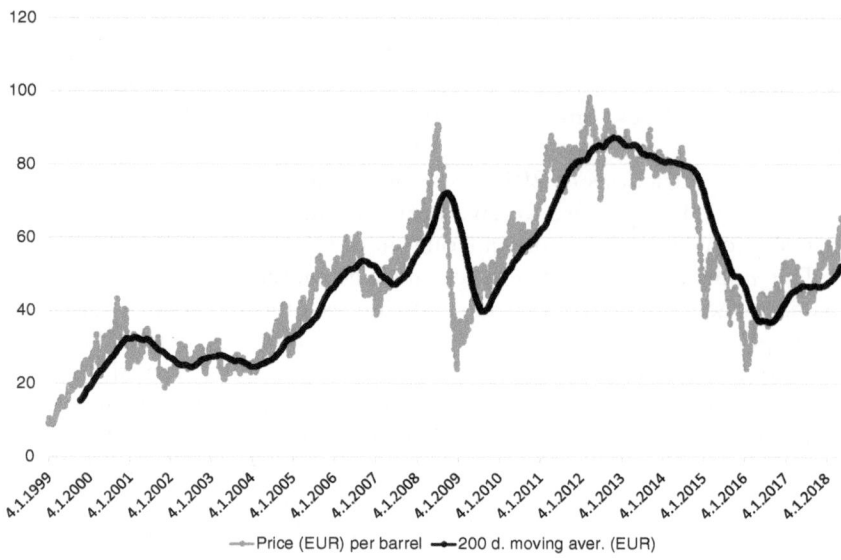

Fig. 2.6 Oil price (Brent quality) in euros during the period of January 1999–May 2018 (*Source (data)* EIA 2018)

March 2012, when barrel of Brent quality cost 98.1 EUR. It should be noted that the peak price was reached in Europe after the 2009 global credit crunch, and even the moving average (200 days) was considerably higher during years 2011–2014 compared with the short-lived spike in 2008. This further clarifies why companies in Europe, and particularly those operating in the Baltic and North Seas, were worried about implementing demanding sulphur level caps during the year 2015.

Lower prices caused by late 2014–2015 oil price correction were shorter-lived in EUR terms compared with USD prices. In 2016, oil prices had already started to clearly stabilize, with lows recorded in August 2016, when the 200-day average was around 36–37 EUR per barrel. After this, prices increased, and the moving average experienced a clear increase in late 2017. Simply observing annual averages could shed further light on this: in 2015, within euro terms, Brent oil was 36.5% lower compared with the yearly average of 2014. In 2016, this decline continued with a 16.2% drop (as compared with year 2015). However, in 2017, average prices increased by 21.4% from the previous year. From the beginning of the year to mid-May 2018, prices have continued to increase and have shown a rise of 18%.

2.5   MARITIME DIESEL: PRICE GOES UP AS LOWER SULPHUR QUALITY REQUIREMENTS RISE

There is a fundamental shift in price with the desire for cleaner maritime types of diesel—the lower the sulphur content level, the higher the price. Of course, prices differ globally based on local area demand and supply, but on average there is a clear premium on the used diesel fuel. Figure 2.7 illustrates this further in the case of the port of Rotterdam, from the last week of February to early March 2018, as well as from another 10-day period in June 2018. On average, very low sulphur content (maximum 0.1%) Marine Gasoil (MGO) cost from 538–640 USD per ton. High sulphur content diesel oil (maximum 3.5%—what world

Fig. 2.7 Rotterdam sea port prices (USD per ton) of MGO (maximum 0.1% sulphur content) and IFO 380 (maximum 3.5% sulphur content) maritime diesel oils from 23 February–8 March 2018 (first period), and 5 June–18 June 2018 (second period) (*Source (data)* Bunker Index 2018)

uses in general until the end of 2019) in turn has the price tag of 339.5–431.5 USD per ton. So, basically this sort of difference will need to be paid if scrubbers are not installed on ships (or if LNG-powered ships are not used). The difference is around 207 USD per ton, approximately a 54% higher fuel cost. This is of course the situation from two very short time periods that took place in 2018, and is affected by local supply and demand, as well as by the general oil price. Do note that this difference is between the globally used sulphur content and the one used in the Baltic Sea, for example. However, the global sulphur content will not be as low as the Baltic Sea, with a maximum of 0.5%. So, the price difference could be a little bit lower.

It is questionable as to whether prices of low-sulphur maritime diesel oil will be kept low enough as compared with higher sulphur content types of diesel. If global maritime supply chains decide to use just MGO (maximum 0.5% sulphur content), then the capacity to produce this grade, and eventually supply it is an open question. In the Baltic Sea during 2015, this costly scenario did not materialize, and supply was available. In addition, oil prices were at low levels during 2015 and 2016, and this aided the situation considerably. Of course, the year 2020 is different, and oil prices could start to climb again, and there may exist global constraints in supply. This could be devastating for the maritime supply chains, and may be a difficult cost to absorb.

It is therefore important to follow global developments in maritime diesel prices, and the difference between lower and higher sulphur content oil. This is somewhat problematic as MGO maritime diesel oil has different sulphur contents around the world. In Fig. 2.8, MGO diesel oil has a maximum of 1.5% sulphur content. The observation time period is from the beginning of 2017 until end of May 2018. It can be clearly seen that MGO prices have increased, from slightly above 600 USD to 765 USD per ton. The relative change from early 2017 to the last observation point at the end of May 2018 is 26.4% (or 159.9 USD in absolute terms). Of course, higher sulphur content diesel oil (IFO 380) also increased in the same period by 34% (or 123.9 USD).

What is most important to notice is that the difference between grades is widening and growing. It shows 14.9% growth in the observation period; using cleaner diesel oil is cost 36 USD more in May 2018 as compared to the beginning of 2017. This is of course harmful information for maritime supply chains in the face of the 2020 sulphur regulation change. However, caution should be applied here given that

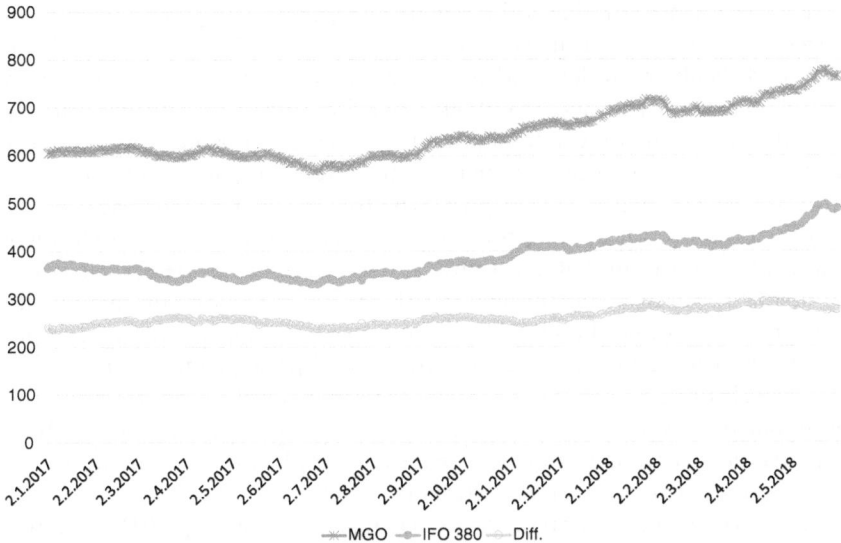

Fig. 2.8 All sea port prices (average) of MGO (maximum 1.5% sulphur content) and IFO 380 (maximum 3.5% sulphur content) maritime diesel oils in 2 January 2017–31 May 2018 (*Source (data)* Bunker Index 2018)

the MGO price is much higher on average around the world as compared with the Rotterdam price, even if world price consists of higher grades. In straightforward terms, it should be lower or around the same. What causes such differences can only be guessed at, but of course, the demand of the Baltic and North Sea for 0.1% sulphur level diesel oil is every day and high. In addition, there is a big oil refinery area in Rotterdam, which produces the necessary grades. So, these factors could have something to do with the illogical price difference.

In relative terms, MGO oil globally has been priced 60–70% higher compared with IFO 380. Even during the observation period of Fig. 2.8, this has varied a lot. At the beginning of 2017, this difference was 66.1%, but increased over a short amount of time and grew to the 77% level in March 2017. Thereafter, the relative price difference has eased and declined. At the end of May 2018, this difference was back at the level of 56.7%. This difference is again slightly higher than what it was in Rotterdam in the same time.

The seriousness of plain oil price changes, as well as grade changes on prices, is harming sea transport in a significant manner (Hilmola 2015). This is difficult to understand for a person without proper background knowledge from the field. Globally, maritime diesel is typically entirely tax-free. If sales include value-added tax (VAT), companies can deduct it in their VAT calculations. Other taxes are often not paid at all. The situation is the same with air transport, which have been able to use tax-free fuel for decades. Of course, these two modes of transport are extremely important for the global economy, and favourable tax treatment could be justified, but as a starting ground this is not good for sulphur quality changes of maritime diesel. Airlines and maritime transportation companies have historically been very sensitive to oil price changes. As another factor is brought to table—namely, very low level sulphur—then the potential downside is serious. This is even the case in situations where negative economic issues do not materialize right away; even the threat can freeze many large-scale investments.

Based on earlier research, it was estimates concerning 2015 sulphur change at the Baltic Sea were such that prices would increase by 5–20% at final service product level (provided transportation service), if oil prices do not change from the year 2014 at all, and the price change in used maritime oil is caused by changing the quality of used maritime diesel (Hilmola 2015). The effect was higher in this study if the ship cost structure was more diesel intensive (old ships and high consuming RoRo/RoPax). However, what should gain attention is the possible significant price increase when oil prices continue to increase in the world markets, and very low level of sulphur content is required to be used. If oil prices increase by 20%, and at the same time, a lower sulphur content is required, then this final service product level price change will be in the range of 8–32%. If oil prices at world markets increase by 50% and sulphur change persists, then the price increase could be anything between 12.5 and 50%.

2.6   Conclusions

All three countries analyzed in this chapter are significantly dependent on international trade. However, trade has developed favourably—mostly in imports. In all three countries, imports were greater than exports at the end of the observation period—in Finland and Sweden imports have also been better sustained than exports (in the case of Finland it has shown a small decline, but in Sweden it has grown more). Only in Estonia is export growth outpacing import development. However, Estonia has been a trade deficit nation for a long time, so the dominance of imports is still strongly present in statistics. All of these countries have remarkably high GDPs (as compared with global levels), and in the observation period Estonia showed remarkable growth from low prosperity levels, being one of the best performers from the old Eastern Bloc in Europe. Of these three countries, Finland and Estonia had problems with trade growth in 2015 (growth in 2016 was also very weak in these countries), but in 2017 they have shown clear and significant growth rates. So, from a macro-economic perspective, sulphur regulation has not brought about big changes since the problems in 2015 could also have originated from the European economic situation and the Ukrainian dispute (and subsequent sanctions). GDPs in 2017 also showed a clear upwards trend in all three countries.

The price of oil is very unpredictable, as the long-term analysis shown in this chapter. The period from the 1990s low to the peak of 2008 was very brief, and the price changed very significantly. However, it should be noted that in euro terms, the highest prices were experienced in early 2012, and not in the year 2008 as per the situation in USD terms. If the measure of the less fluctuating 200-days moving average is used, then in both currencies the highest prices were paid during 2011–2012 (this high average price continued until 2014). Due to the oil price decline in 2014–2015, prices have not in increased or changed that much in terms of the decade time scale. Fluctuation has remained as high. In. addition, moving averages for the future show that price has again upwards momentum.

Oil price change is one harmful factor for physical supply chains, but an additional harmful factor is the higher price of lower sulphur content diesel oil. Statistics clearly show that a price difference exists between higher-emitting grades compared with lower-emitting grades. At the Baltic Sea during early 2018, this difference showed that lower sulphur content diesel was somewhere around 50% more expensive compared with higher sulphur content diesel. This difference is the same globally, and is actually a little bit higher if statistics include all sea ports reporting in the used dataset, and using quite much higher sulphur content, which is labeled as low sulphur maritime diesel. Together with sudden oil price swings upwards, this element adds unfortunate uncertainty to maritime supply chains from 2020 and onwards.

REFERENCES

Bunker Index. (2018). *Bix bunker index, prices.* Available at http://www.bunkerindex.com/prices/index.php. Retrieved 9 March 2018 and 19 June 2018.

EIA. (2018). *Europe Brent spot price FOB (dollars per barrel).* Available at http://tonto.eia.gov/dnav/pet/hist/LeafHandler.ashx?n=PET&s=R-BRTE&f=D. Retrieved 22 May 2018.

Finnish Customs. (2018). *Imports, exports and trade balance in 1970–2017.* Available at http://tulli.fi/tilastot/taulukot/aikasarjat. Retrieved 5 March 2018.

Hilmola, O.-P. (2015). Shipping sulphur regulation, freight transportation prices and diesel markets in the Baltic Sea region. *International Journal of Energy Sector Management, 9*(1), 120–132.

Ministry of Transport and Communications. (2014). *Fairway dues be halved and track tax removed for 2015–2017.* Press release 18 September 2014, Helsinki, Finland. Available at https://www.lvm.fi/en/-/fairway-dues-to-be-halved-and-track-tax-removed-for-2015-2017-795181. Retrieved 22 May 2018.

OECD. (2016). Share of international trade in GDP. In OECD factbook 2015–2016: Economic, environmental and social statistics. Paris: OECD Publishing. http://dx.doi.org/10.1787/factbook-2015-25-en.

Statistics Estonia. (2018). *Statistical database: Economy.* Available at http://pub.stat.ee/px-web.2001/I_Databas/Economy/databasetree.asp. Retrieved 5 March 2018.

Statistics Finland. (2018). *Statistical database: Annual national accounts.* Available at http://pxnet2.stat.fi/PXWeb/pxweb/en/StatFin/StatFin__kan__vtp/?rxid=789d15f2-484d-47f3-a0c1-d8438204d7ab. Retrieved 5 March 2018.

Statistics Sweden. (2018). *Statistical database: National accounts and trade in goods and services.* Available at http://www.statistikdatabasen.scb.se/pxweb/en/ssd/. Retrieved 5 March 2018.

Stiglitz, J. E. (2018). *The euro: How common currency threatens the future of Europe* (1st ed.). London, UK: W. W. Norton.

CHAPTER 3

Was Sulphur Regulation the Reason for Growth of Unitized Cargo Between Finland and Estonia?

Abstract Within the European map, Finland is a little like an island; the country is in peninsula position within the northern Baltic Sea. Therefore, foreign transportation flows are tied to the sea. In the Baltic Sea region, Finland's most important foreign destinations of unitized cargo (containers, semi-trailers and trucks) are the sea ports of Germany, Sweden and Estonia. Previously, Germany and Sweden were clearly dominating in terms of cargo flows. However, Estonia has continuously grown (and continues to do so) and in 2015 it reached the same levels of volumes with Sweden. There are number of reasons and long-term drivers that make the Estonian route attractive, but its ability to match Swedish volumes in 2015 was mostly enabled by changes in environmental legislation. During 2010 and 2015, the Baltic Sea region faced sulphur regulation changes with companies required to use shipping fuel with a much lower sulphur content. In regression models, both of these years are significant in explaining freight volume change (growth in Estonia and decline in Sweden), however, 2015 in particular was clearly a game changer (with decline even in the German route).

Keywords Unitized cargo · Finland · Estonia · Baltic Sea · Sulphur regulation

3.1   INTRODUCTION

In recent years, one of the most significant change in the logistics sector for the Baltic Sea has arguably been implementation of sulphur regulation in 2010 and 2015 (IMO 2018). In particular, the latter steep cut in the sulphur content of maritime diesel oil was considered to be risky and costly (Kalli et al. 2009; Delhaye et al. 2010; Notteboom 2011; Hilmola 2015; Zis and Psaraftis 2017). However, the future is always difficult to forecast. This was the case here too as oil markets started to tank and the price of oil lost considerable value in the years 2014–2015 (Hämäläinen et al. 2016). Therefore, expected significant changes in 2015 were actually minor few percent increases in sea transportation costs based on respondents from Estonia, Finland and Sweden (Hilmola et al. 2017). Similar very small scale effects of 2015 have been reported from southern parts of the Baltic Sea as well as the North Sea area (Zis and Psaraftis 2017). A large-scale logistical cost survey in Finland revealed that in 2015, overall costs of both logistics (+3.7%) and transportation (+20.5%) increased from the previous observation year (2013) (Solakivi et al. 2017). Even if this change could be considered, based on these figures, to be a minor or moderate change, it was actually rather significant. At the same time in 2015, and in the following years, shipping globally provided very low prices, as well as declining prices (United Nations 2017). So, in this vein, the change was negative, and the negative effects of very demanding sulphur regulations are still present in Baltic Sea shipping. Many companies are now investing in Liquefied Natural Gas (LNG) technology and LNG-powered ships (in order to avoid the harm from these emissions as diesel prices increased substantially in 2017, as well as to tackle future demands for nitrogen emissions and CO_2). The cost of these new LNG ships are much higher than conventional diesel-powered shops, and there does not exist outsight that lower freight rates would be reality in Baltic Sea area. In fact, it quite the opposite.

It is very tempting to complete freight flow analysis from the change caused during sulphur regulation years to northern part of Baltic Sea. At the time of writing of this work, there are available data concerning the years 2015–2017, which makes it uniquely possible to analyze the effects (especially in the latter part of 2015) of implementing the demanding sulphur regulations. The reason is that this area is European periphery and in many studies it has been considered to be the main party carrying the liability for costs, while the benefits are experienced in areas of Europe that are more populated. In addition, the global sulphur regulation level of 2020 is very close (IMO 2018), and similar changes could be expected in global shipping routes, what have been taken place in analyzed transportation system of this chapter, unitized cargo (containers, semi-trailers, trucks etc.) foreign flows of Finland. In Finland, for foreign trade, sea transport is a vital component, and within the same magnitude (having a share of 83% in 2015; Finnish Customs 2016), as for global trade (80% based on United Nations 2017). From Finland, shorter and longer unitized cargo routes exist for sea transport, similar to the options for transport containers using deep sea ships to Asia (such as through the Suez canal or around Africa). The research questions of this chapter are the following: could support be found from long-term statistics that sulphur regulation has had its effects on the logistics flows of Finland? What kind of effects have there been for shorter routes, and by contrast, for longer routes? This work mostly uses long-term macro-data from the years 1998–2017.

This research is structured as follows: In Sect. 3.2 earlier research regarding sulphur regulation implementation and logistics sector change is reviewed. The research environment follows thereafter in Sect. 3.3, where are also three main shipping routes introduced and their development in the years 1998–2017 is analyzed based on the time-series. Empirical data analysis using regression model is completed in Sect. 3.4. Finally, in Sect. 3.5 the study is concluded, discussed, and future research avenues are proposed.

3.2 Sulphur Regulation and Logistics Sector Change

The European Union has for a long time been at the forefront of environmental issues and tackling climate change. It was already roughly two decades ago when the first demands for sulphur regulation in shipping became a reality (Council Directive 1999/32/EC). The Baltic Sea and North Sea were already at a demanding level in 2005–2006, when the sulphur content of diesel fuels within ships was lowered to 1.5%. This lowering trend continued in 2010 as Sulphur content was placed at 1.0% based on the initial agreed programme. The entire world of course followed, but from a very long distance as sulphur content was lowered from 4.5% to 3.5% in year 2012 (IMO 2018). From 2015 onwards, in the Baltic Sea and North Sea area, the sulphur limit has been as low as 0.1% (IMO 2018). The entire world will implement a level of 0.5% in year 2020 (if the current decision holds). Sulphur regulation is typical example of environmental legislation that is not the same for the entire world and actually creates areas where emissions are more greatly penalized. This will of course lead to global inefficiencies and higher costs with regard to the desired outcomes—Tol (2010) argues that "global" and "same for all" policy is the most effective in combating environmental harms.

These strict demands are of course workable and beneficial for Central Europe, which has one of the world's largest population densities with respect to geographical area. As this area has major shipping routes outside of the Baltic Sea, new demands did not lead to excessive or indeed any cost increases at all, but the societal benefits have been rather substantial (due to lives saved, and lower numbers of illnesses and allergies; Antturi et al. 2016; Lähteenmäki-Uutela et al. 2017). Countries in the European periphery will of course suffer (Dannenberg et al. 2007; Antturi et al. 2016) from higher demands as these countries will end up paying higher costs (or be at a cost disadvantage as compared with other regions and countries), and disruptions and reconfigurations take place in supply chains. Higher costs could also prevent industrial and logistics sector job loss in the future as units are placed in more advantageous locations. Of course, the situation is not that bleak since countries in general, and the technology export companies of economy in particular, can benefit from early implementations in their country and areas in close proximity (Lähteenmäki-Uutela et al. 2017). As in the case of sulphur regulation, the world will demand new technologies and solutions due to the changes of 2020, and these are already available in areas and companies that have faced these demands before (and in stricter forms).

Earlier research on this topic has agreed on one thing and that is the fact that higher demands will lead to higher costs. Strategies to tackle this differ. Shipping company may invest in scrubbers and continue to use very high sulphur content diesel oil (Ma et al. 2012; Yang et al. 2012), or alternatively just fine-tune motors and start to use 0.1% level sulphur diesel, for example (Zis et al. 2015). Both of these will cost—the first will require millions of euros investment and ensuring new machinery is operational (e.g., for unloading sulphur from the machine, the necessary maintenance and operational electricity), while the latter just leads to higher fuel cost because lower sulphur content diesel has a higher price tag. Both of these will lead to a situation in which longer shipping routes in short-sea shipping will be hurt, and shorter routes will benefit.

It is question of whether it is sensible to replace sea transportation with hinterland modes (road and/or railways). In the case that distances do not increase that much, it clearly benefits very short distance sea journeys at the expense of longer journeys. In periphery (distant and lower population), it is often so that unitized cargo is either transported in RoPax (where passengers enjoy sea journeys together with cargo on the ship, in terms of trucks and semi-trailers) or RoRo ships. RoPax and RoRo ships consume much more diesel fuel than other ships, such as containers or bulk, and therefore these are mostly economically hurt. This excessive cost is not realized to a great extent in the short-distance journeys, but more so in longer distance journeys.

In the Baltic Sea area, it was stated that sea journeys, for example, between Finland and Estonia (Hilmola 2014) as well as between Sweden and Germany were to benefit whatever the situation may be after implementation. These are very short-distance sea routes. In addition, hinterland transport from Sweden, or through the Baltic States (Notteboom 2011), should considerably benefit since diesel fuel tax treatment of hinterland transport mode is already fixed (at sea, fuel is typically tax free) and basically sulphur free. The change will not concern hinterland transportation, since this will only be more competitive compared with sea (Hilmola 2015). However, road usage payments could erode the situation somewhat. Changes did not materialize to their full extent as oil price suddenly dropped in the late 2014 and 2015. This was caused by the Ukrainian dispute, and was also a result of the global economic problems, together with increased USA-based shale oil production. It was so that fuel prices ultimately increased rather moderately in 2015 within the ships operating in the Baltic Sea area (Zis and Psaraftis 2017; Hilmola et al. 2017). Many companies that decided to invest in scrubbers in 2014–2015 actually made a badly timed investment as they probably should have invested in these systems in 2017 or later (as oil prices have recovered to higher levels). In addition, later investment would have hedged against technological uncertainty and technical errors. Companies could have also used the extra time to rethink what sort of fleet and with what scale would best suit this new situation. It could be concluded from the implementation of 2015 sulphur regulation that it caused harm and did not lower the freight rates (since elsewhere companies were enjoying very low rates; United Nations 2017); however, the worst-case scenarios were not realized.

3.3   RESEARCH ENVIRONMENT: FINNISH UNITIZED CARGO
TRANSPORTATION VOLUMES AT THE BALTIC SEA

Most of the Finnish foreign trade is implemented through Baltic Sea
transports. In short sea shipping routes, countries like Germany, Sweden
and Estonia play a vital role (Fig. 3.1). In the late 1990s and early 2000
it was crystal clear that Germany and Sweden were the destinations for
unitized cargo, since from the total volume of these three countries
Germany and Sweden took more than 90%. Estonia was really marginal
in unitized freight back then, but constant growth changed everything,
and the marginal route became the mainstream. Currently, the share
of Germany and Sweden is somewhere above three fourths of the volume, and Estonia has a share of nearly one fourth. In revisiting Fig. 3.1,
it could be detected that German and Swedish transportation volumes
linked to/from Finland do not show that much growth, and actually,
after 2007, they have been on declining trend. Estonia has in turn constantly grown. If these three markets are summed together, volumes of

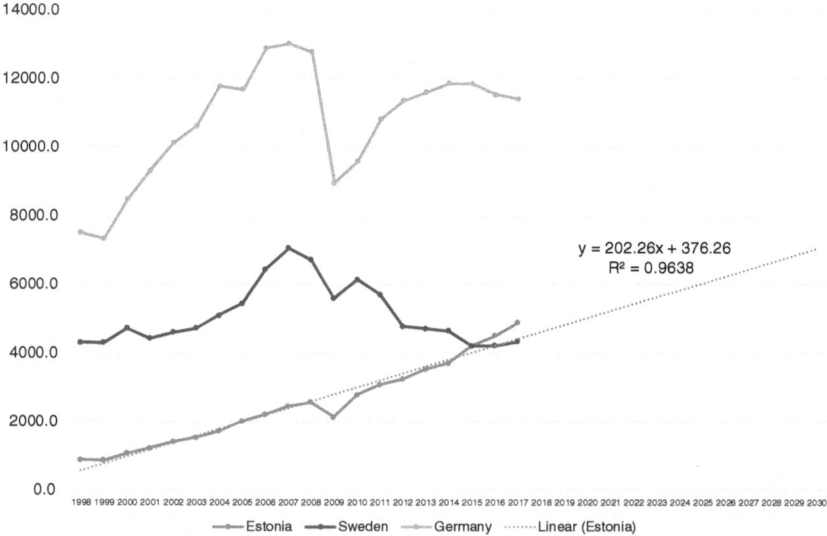

Fig. 3.1 Unitized cargo volume development to and from Finland in the years
1998–2017

the most recent years are around the level of the years 2006–2007. So, growth has been very minor in total after slump in demand of 2009.

Of course, it could be argued that German and Swedish trade with Finland is the cause for this change. However, in the long-term and also during years 1998–2017 trade was growing very nicely: German trade in year 2017 was double in euro terms compared with 1998, and Swedish trade was in turn 1.84 times higher. Estonian trade roughly doubled in this period. It should be emphasized that the absolute size difference in trade volumes is significant. Swedish trade is 3.6 times higher than Estonian trade (for the year 2017), and German trade is correspondingly nearly five times higher. So, most of the Estonian growth arises from transit traffic to other Baltic States, Poland, and even to destinations farther away like Czech Republic, Slovakia, Hungary, Romania and Turkey. On the other side, Sweden has been losing transit traffic, as has Germany. One reason could be the increased cost level in terms of salaries, sea port payments, but also environmental payments and restrictions on roads. Logistics flows are always like water, seeking the most convenient, proper quality and cost competitive routes.

In the observation period of 1998–2017, the average annual growth was nearly 9.4% in Estonia and it was close to 0% in Sweden, and 2.2% in Germany. This is of course the average annual growth—if the only good years for Germany and Sweden were taken into account (1998–2007), both of these countries would show growth of around 6%. Estonia back then was growing at 12% p.a., but its starting level was really low. The biggest change took place in years 2008–2017. Sweden showed annual decline of 4.7%, Germany showed a decline of 1.3%, but the Estonian route grew on average 7.4% p.a. It should be noted that year 2009 was really difficult for all routes, and Estonia as well as Sweden suffered around a 16.5% decline, and Germany a decline of 30%.

Sulphur regulation has been implemented in a step-wise manner within the Baltic Sea, with 2010 and 2015 being the most important years (the initial implementation of 1.5% level in 2005–2006 was not that costly). Interestingly, in Fig. 3.1, these years were very strong years for Estonian cargo volume growth. Even if the Estonian route experienced severe difficulties in year 2009, it grew so much in 2010 that cargo volume levels of 2008 were already passed back. The same cannot be said for German or Swedish cargo volume recoveries in the year 2010—they were particularly weak, especially in the case of Germany. This weakness could have been caused by trade weakness between countries or by lack of competition between transit services. Therefore, and to tackle these possible weaknesses, in the following regression analysis models contain trade volumes (in EUR) as one explanatory variable.

The year 2015 was the time of implementation for stiff sulphur regulation (maximum 0.1%). Estonian cargo volumes subsequently grew by 13.7%, while Swedish cargo volumes suffered a 9.4% fall, and German cargo volumes very slightly declined. It is rather apparent that sulphur regulation led to further growth for the Estonian cargo volumes, as can be seen from Fig. 3.1, when long-term linear trend is compared to the actual volume. In 2015, Estonian cargo volumes matched those of Sweden and in 2016, they were already higher. This situation has continued in 2017.

3.4   EMPIRICAL DATA ANALYSIS: FINNISH REGRESSION MODELS

3.4.1   Estonian Sea Transportation Flows

From Finnish perspective, Estonia is nowadays often the preferred short sea shipping destination since the service level is high (due to the number of daily RoPax connections), transportation lead times are short, sea transportation is accurate regardless of the season, and transportation is not affected by uncertainty of strikes (typically in Finland sea ports are labour strike sensitive, however, these events will not affect RoPax as trucks are just driven by non-unionized and usually foreign drivers inside the vessel). Most of the cargo volume on the unitized freight side is between the sea ports of Helsinki and Tallinn (above 85%). The rest of the volume is basically between Hanko and Paldiski. Some limited container transport volume is from named Estonian sea ports to HaminaKotka, Helsinki, Hanko and Rauma. The maritime transport distance between Estonia and Finland of course depends on the sea ports, but it is typically around 90–100 km. The distance from Tallinn to HaminaKotka is around 180 km, and to Rauma it I saround 500 km. So, most of these are very short or short shipping routes.

To explain unitized cargo development between Finland and Estonia the following parameters were used: total trade in euros between Finland and Estonia; time (year); EU/Nato membership (binary); sulphur regulation of year 2010 (binary); and sulphur regulation of the year 2015 (binary). The first binary explaining variable was the EU/Nato membership, which was zero before year 2004 and thereafter was one. In terms of sulphur regulation, possible changes were modelled with binary numbers: The year 2010 change was one during years 2010–2014, and year 2015 demanding implementation being one during years 2015–2017 (in other years zero). All of the data for the regression model was gained from national statistics and was personally gathered from different sources over the years. The examination period covered years 1998–2017 (20 years of observation).

Within the Estonian regression model, it was found that time itself is the explanatory variable with high statistical significance; in other words, cargo volume growth occurs in a time progression basis and every year unitized cargo volume grows by 0.137 million tons (see Table 3.1 for details). In addition, total trade had a positive co-efficient and was clearly statistically significant. Sulphur regulation changes in the years 2010 and 2015 had positive impacts on the Estonian route. During 2010, the impact was a little bit small, somewhere below 0.2 million tons, but was nevertheless still significant. However, the implementation of sulphur regulation in 2015 brought approximately 0.8 million tons more unitized cargo on this route. If these effects are presented in semi-trailer or forty-foot equivalent unit (FEU) basis, 0.189 million tons corresponds to around 12,179–13,019 units per year (weight is 14.5–15.5 tons per unit), and 0.824 million tons in turn corresponds to 53,153–56,819 units per year. It is not of course the case that the total amount for the years 2010 and 2015 would have been added to the route in total, because 2010 change was zero after 2015 (as sulphur regulation changes were modelled within binary fashion). The current effect can be estimated at around 0.8 million tons.

The regression model itself provides really high explanatory power with these variables since R^2 is 99.5% (Table 3.1). The standard error is also very low, below 0.1 million tons. So, for the years 1998–2017, this model forecasts cargo volumes really well. However, it should be noted with caution that models are never sure forecasters of the future—the logistics system could experience changes such that it is no longer accurate or the effect diminishes. For example, Hilmola (2014) built regression models for Finnish–Estonian unitized cargo with data from the years 1998–2013 (four years fewer compared with this case). Sulphur regulation was not included in these models. EU/Nato membership was clearly positive in cargo forecasting models back then; however, now, as sulphur regulation is properly included, its membership effects are not statistically significant. The problem is of course the situation where forecasted variables are growing all the time. Numerous phenomena and variables could be argued to have significance.

Table 3.1 Regression model to explain unitized cargo transport between Finland and Estonia from 1998–2017 (cargo amounts in thousand tons)

SUMMARY OUTPUT

Regression statistics

Multiple R	0.997
R square	0.995
Adjusted R square	0.994
Standard error	97.860
Observations	20

ANOVA

	df	SS	MS	F	Significance F
Regression	4	28083534.1	7020883.5	733.1	5.33608E−17
Residual	15	143648.8	9576.6		
Total	19	28227182.8			

	Coefficients	Standard error	t Stat	p-value	Lower 95%	Upper 95%
Intercept	−2.7363058E+05	1.9143346E+04	−1.4293770E+01	3.8274203E−10	−3.1443366E+05	−2.3282751E+05
Year	136.9909	9.6000	14.2699	0.0000	116.5290	157.4527
Total trade	3.53169E−07	6.61697E−08	5.337314924	8.29612E−05	2.12131E−07	4.94206E−07
Sulphur (0.1%)	823.8697	117.1020	7.0355	0.0000	574.2727	1073.4667
Sulphur (1.0%)	188.7683	84.7514	2.2273	0.0417	8.1250	369.4117

3.4.2   Swedish Sea Transportation Flows

The main routes between Finland and Sweden in unitized cargo are those operating between Naantali and Kapellskär (RoRo) as well as Turku and Stockholm (RoPax). The third most popular is between Helsinki and Stockholm (RoPax). These three routes take most of the cargo volumes. In Finland, at sea port basis Naantali, Turku and Helsinki account for around 90% of overall volume (two first mentioned make up 75%). Other sea ports also contribute to connections, but in cargo volume terms they are not that significant. In the north there are RoPax connection between the cities of Vaasa and Umeå (RoPax), and some container sea ports of Finland (like Oulu and Tornio) have sparse volumes to Göteborg. The distances between sea ports are not extremely long; the shortest routes start at 100 km (Vaasa–Umeå) and increase to somewhere above 200 km (Naantali–Kapellskär). They continue from there to 500 km and the longest is approximately 2000 km between Tornio and Göteborg. The shortest routes are typically popular for cargo transport.

The variables explaining unitized cargo volume to Sweden were similar to those in the earlier model (EU member already since 1995, while not being part of military alliance, Nato): total trade in euros between Finland and Sweden; time (year); sulphur regulation of year 2010 (binary); and sulphur regulation of year 2015 (binary).

What was interesting in the Finnish-Swedish regression model concerning unitized cargo transport was the lack of a time component in the final model (it was not even close to being statistically significant; see Table 3.2 for details). Only total trade between countries is considered to be clearly statistically significant, showing a positive co-efficient, and driving growth between Sweden and Finland. The trade co-efficient is much higher in the Swedish model than was presented earlier in the Estonian model, so freight volume is more sensitive to direct euro trade amount changes (please note that Swedish flows regression model does not include fixed intercept like the Estonian model did). Since the time component was lacking in the model, sulphur regulation effects were considered to be very negative in both 2010 and 2015, and they were both statistically clearly significant. Based on the regression model in 2010, sulphur regulation implementation in the Swedish route lost 1.67 million tons of unitized cargo transport. In turn, implementation of sulphur regulation in 2015 resulted in the loss of 2.4 million tons. If these huge amounts are converted to forty-foot equivalent units, then 1.67 million tons corresponds

Table 3.2 Regression model to explain unitized cargo transport between Finland and Sweden from 1998–2017 (cargo amounts in thousand tons)

SUMMARY OUTPUT

Regression statistics

Multiple R	0.994
R square	0.989
Adjusted R square	0.929
Standard error	589.094
Observations	20

ANOVA

	df	SS	MS	F	Significance F
Regression	3	5309969405.4	176989801.8	510.0	4.3727E−16
Residual	17	5899531.9	347031.3		
Total	20	536868937.3			

	Coefficients	Standard error	t Stat	p-value	Lower 95%	Upper 95%
Total trade	5.48125E−07	1.74681E−08	31.37862042	1.72E−16	5.11271E−07	5.8498E−07
Sulphur (0.1%)	−2398.549	400.784	−5.985	0.000	−3244.129	−1552.968
Sulphur (1.0%)	−1673.408	342.533	−4.885	0.000	−2396.090	−950.725

107,962–115,407 units per year (weight is 14.5–15.5 tons per unit), and 2.4 million tons corresponds to 154,745–165,417 units per year. However, it should be emphasized that the loss is around 2.4 million tons, as the effects of the year 2010 were not accounted for in the model during the last three years.

The Swedish cargo regression model also shows high explanatory power with these variables as the R^2 value is 98.9% (Table 3.2). As a note of caution, it should be stated that the standard error is rather high, at 0.589 million tons. Therefore, the above-stated effects of sulphur regulation could be overestimated in the long-term. However, this does not change the fact that sulphur regulation clearly had very negative effects on the Swedish route, but these values may be too extreme since model is built with limited years (20) in it.

3.4.3   German Sea Transportation Flows

The longest distance routes from Finland at the Baltic Sea area end at Germany. In container transportation branch feeder ships serve the hub sea ports of Hamburg, Bremerhaven and Wilhelmshaven (these are located at shores of the North Sea, but most of the journey is through the Baltic Sea), for which the most important connecting points are in Finnish side, in Helsinki, HaminaKotka and Rauma. Other unitized cargo connections are RoRo based at German sea ports such as Travemünde, Rostok and Lybeck, which are served through the Finnish sea ports of Helsinki and Hanko. German unitized cargo transportation is also rather concentrated at a few sea ports; Helsinki, HaminaKotka, Hanko, and Rauma account for around 80% of the overall volume (Helsinki and HaminaKotka make up approximately 50%). Distances to German sea ports are great since typical RoRo sea ports require a sea journey of around 1000 km, while container-based transportation (going around Denmark) requires nearly 2000 km.

The German regression model of unitized cargo shared similarities with the earlier two models as it included the following possible parameters for statistical testing: total trade in euros between Finland and Germany; time (year); sulphur regulation of year 2010 (binary); and sulphur regulation of year 2015 (binary).

Regression modelling ended with a model in which the results were similar to the Swedish model. So, time was not considered to be a statistically significant variable for explaining cargo volumes on this route (Table 3.3). The same applies to sulphur regulation changes in 2010 since this was not statistically significant. However, sulphur regulation change in 2015 really affected the volumes as it was considered to have removed 2.38 million tons from this route. Total trade is of course positive and a statistically significant driver for the unitized cargo here. It clearly has the highest co-efficient as compared with Swedish and Estonian regression model (please note that regression model of German flows does not include fixed intercept). This means that German unitized cargo is more dependent on bilateral trade between Germany and Finland.

Again, the explanatory power of this built regression model for unitized cargo volumes is high since the R^2 value is 99.2% (Table 3.3). However, the weakness of the model is its standard error—nearly one million tons.

3.5   Concluding Discussion

Environmental demands are already high for shipping at the Baltic Sea. However, it is obvious that transportation and logistics shall face continued pressure from different emission prevention plans. This sector is great contributor to CO_2 emissions alone, and these need to be further reduced in the following decade in European Union area (especially in Finland and Sweden). After implementation of sulphur regulation, there has been a lot of talk about implementing nitrogen legislation in the Baltic Sea area. Based on the latest decision of IMO in this regard, the Baltic Sea and North Sea will face a situation in 2021 where all new ships will either have some technical mechanism to clean diesel (catalytic) or alternatively they will use only LNG (Ministry of Transport and Communications 2016).

This research finds that environmental demands (e.g., implementing stiff sulphur regulations) clearly impact on logistics routes, and in the case of Finland, routes were more hinterland-based using only very short sea journeys, like shorter Estonian ones. This finding is interesting, not only for the Baltic Sea area, but also for the wider global implementation of low sulphur levels for shipping during the year 2020. Whatever the development of oil price in 2020 and in the years thereafter, this change will most probably mean that shorter routes will be favoured globally for logistics. In continental transports, it could mean that growth in sea ports occurs in the south of Europe where Asian containers are increasingly transported by hinterlands to their final consumption. This has already taken place in the sea ports of Koper (Croatia), Piraeus (Greece) and

Table 3.3   Regression model to explain unitized cargo transport between Finland and Germany from 1998–2017 (cargo amounts in thousand tons)

SUMMARY OUTPUT

Regression statistics

Multiple R	0.996
R square	0.993
Adjusted R square	0.937
Standard error	983.222
Observations	20

ANOVA

	df	SS	MS	F	Significance F
Regression	2	2335249749.1	1167624874.5	1207.8	4.75522E−19
Residual	18	17401055.4	966725.3		
Total	20	2352650804.4			

	Coefficients	Standard error	t Stat	p-value	Lower 95%	Upper 95%
Total trade	8.57585E−07	1.91832E−08	44.7049727	6.69928E−20	8.17283E−07	8.97887E−07
Sulphur (0.1%)	−2376.906	647.995	−3.668	0.002	−3738.292	−1015.520

some Spanish and Italian sea ports. Transportation between Europe and the Middle East and Europe and Asia shall have increased hinterland transports as an option (since it is shorter and based on railways). In addition, over-capacity and cost emphasis of very low speeds have attracted some container shipping lines to go around Africa in order to reach Asia. This may be seen to shrink in the 2020 shipping world (if geopolitical factors are not taken into account).

For the Baltic Sea area, the findings of this study are vital for the future development of short sea shipping. Now, the key issue seems to be for companies to invest in new expensive ships, preferably LNG. However, it is questionable whether this will happen and whether it will be realized at a large-scale. Another option is for shippers to continue to invest in new cleaners and technologies to get rid of sulphur and, in the future, nitrogen. CO_2 emissions will be a concern in the following decades and in this way LNG again holds the key to continuity and for the solution of shipping operations. Should there be some constraints in the LNG use, it will mean that shorter sea routes and hinterland transport will take the market share from short sea shipping.

REFERENCES

Antturi, J., Hänninen, O., Jalkanen, J.-P., Johansson, L., Prank, M., Sofiev, M., et al. (2016). Costs and benefits of low-sulphur fuel standard for Baltic Sea shipping. *Journal of Environmental Management, 184,* 431–440.

Dannenberg, A., Tim, M., & Ulf, M. (2007). *What does Europe pay for clean energy? Review of macroeconomic simulation studies* (ZEW Discussion Papers, No. 7-19). Mannheim.

Delhaye, E., Breemersch, T., Vanherle, K., Kehoe, J., Liddane, M., & Riordan, K. (2010). *COMPASS—The competitiveness of European short-sea freight shipping compared with road and rail transport.* Leuven, Belgium: Transport & Mobility Leuven.

Finnish Customs. (2016). *Ulkomaankaupan kuljetukset 2015 (in Finnish, free translation to English: "Foreign trade transports of 2015").* Helsinki, Finland: Finnish Customs.

Hämäläinen, E., Hilmola, O.-P., & Tolli, A. (2016). North European export industry and the shadows of sulphur directive. *Journal of Transport and Telecommunication, 17*(1), 9–17.

Hilmola, O.-P. (2014). Growth drivers of Finnish-Estonian general cargo transports. *Fennia—International Journal of Geography, 192*(2), 100–119.

Hilmola, O.-P. (2015). Shipping sulphur regulation, freight transportation prices and diesel markets in the Baltic Sea region. *International Journal of Energy Sector Management, 9*(1), 120–132.

Hilmola, O.-P., Kiisler, A., & Hilletofth, P. (2017). Cabotage and sulphur regulation change: Cost effects to Northern Europe. *International Journal of Business and Systems Research, 11*(4), 417–428.

IMO. (2018). *Sulphur oxides (SOx) and particulate matter (PM)—regulation 14.* Available at http://www.imo.org/en/OurWork/Environment/ PollutionPrevention/AirPollution/Pages/Sulphur-oxides-(SOx)- %E2%80%93-Regulation-14.aspx. Retrieved 19 February 2018.

Kalli, J., Karvonen, T., & Makkonen, T. (2009). *Sulphur content in ships bunker fuel in 2015: A study on the impacts of the new IMO regulations and transportation costs.* Ministry of Transport and Communications, No. 31, Helsinki, Finland.

Lähteenmäki-Uutela, A., Repka, S., Haukioja, T., & Pohjola, T. (2017). How to recognize and measure the economic impacts of environmental regulation: The sulphur emission control area case. *Journal of Cleaner Production, 154,* 553–565.

Ma, H., Koen, S., Xavier, R. P., & Nigel, T. (2012). Well-to-wake energy and greenhouse gas analysis of Sox abatement options for the marine industry. *Transportation Research Part D, 17*(7), 301–308.

Ministry of Transport and Communications. (2016). *Nitrogen emissions from ships restricted in the Baltic Sea and North Sea.* Press release 28 October 2016, Helsinki, Finland. Available at https://www.lvm.fi/-/nitrogen-emissions-from-ships-restricted-in-the-baltic-sea-and-north-sea. Retrieved 20 February 2018.

Notteboom, T. (2011). The impact of low sulphur fuel requirements in shipping on the competitiveness of roro shipping in Northern Europe. *WMU Journal of Maritime Affairs, 10*(1, April), 63–95.

Solakivi, T., Ojala, L., Laari, S., Lorentz, H., Töyli, J., Malmsten, J., Lehtinen, N., & Ojala, E. (2017). *Finland state of logistics 2016.* Publications of Turku School of Economics (University of Turku), E1:2017. Turku, Finland.

Tol, R. S. J. (2010). Carbon dioxide mitigation. In Lomborg, (Ed.), *Smart solutions to climate change.* Cambridge, MA: Cambridge University Press.

United Nations. (2017). *Review of maritime transport.* Geneva: Unctad, United Nations.

Yang, Z. I., Zhang, D., Caglayan, O., Jenkinson, I. D., Bonsall, S., Wang, J., et al. (2012). Selection of technologies for reducing shipping Nox and Sox emissions. *Transportation Research Part D, 17*(7), 478–486.

Zis, T., North, R. J., Angeloudis, P., Ochieng, W. Y., & Bell, M. G. H. (2015). The environmental balance of shipping emissions reduction strategies. *Transportation Research Record: Journal of the Transportation Research Board, 2479,* 25–33.

Zis, T., & Psaraftis, H. N. (2017). The implications of the new sulphur limits on the European Ro-Ro sector. *Transportation Research Part D, 52,* 185–201.

CHAPTER 4

Unitized Cargo: Growing Truck-Based Volumes at the Sea Ports of Estonia, Sweden and Finland

Abstract Using trucks in hinterland transportation has a long tradition in Europe. Typically, in a cross-border context, trucks accompanied by a semi-trailer unit are the most commonly used. All northern Baltic Sea countries, such as Finland, Sweden and the three Baltic States rely on these as part of large-scale in European supply chains (after sea port handling volumes): (in tons) or international railway volumes hardly grew. However, trucks accompanied by semi-trailer units have continued to grow. This is mostly seen in the number of truck handling volumes at the sea ports of Estonia, Sweden and Finland. Growth has been strong for the truck and trailer combination—this means that shorter sea journeys have been favoured. In the long-term, trucking units show even better growth than containers, which have experienced some growing pains.

Keywords Unitized cargo · Trucks · Estonia · Sweden · Finland

4.1 Introduction

The European Union is at the moment (with the UK still a member) the largest economic area in the world, roughly equal to the USA. This is the case with all GDPs of member countries combined together. It was larger than USA prior to year 2015 as there were no disputes and sanctions with the east—namely Russia. The magnitude of the European economy and its influence on the global economy is typically

© The Author(s) 2019
O.-P. Hilmola, *The Sulphur Cap in Maritime Supply Chains*,
https://doi.org/10.1007/978-3-319-98545-9_4

forgotten—it is not taken sufficiently seriously and with enough weight in various analyses. European logistics solutions heavily depend on EU legislation and its demands to member countries. For a long time, this economic area has tried to act proactively, and be at the forefront of environmental issues; different measures as well as restrictions have been implemented in order to support ambitious goals. In one way, it could be seen as testing ground for the rest of the world—for example, with implementation of emission trading, road use payments, taxes of road transportation vehicles, emission-based taxation of vehicles, harmonized taxation of energy products, and different environmental restrictions for used fuels. One of these measures to tackle the worsening climate quality (Particulate Matter PM) was the implementation of sulphur directive at the maritime sector. Basically, the North Sea and Baltic Sea have become testing grounds, where strict demands were implemented in 2010 (maximum 1.0% sulphur content on maritime diesel), and in 2015 (maximum 0.1% sulphur content). In 2020, the entire world should absorb a sulphur level of maximum 0.5% in the maritime sector (down from as high as a maximum of 3.5%).

The European logistics system differs from that of Asia or North America. Typically, in Asia, containers are favoured as sea transportation is the major component in these supply chains. Using containers is the cheapest and most convenient option because hinterland access is often troublesome and the quality of roads and railways vary. The situation is similar in North America, where railways are used more in hinterlands, and therefore it is natural to combine these with seamlessly with sea ports with containers. Of course, in North American intermodal systems, semi-trailers have their place. However, the situation in Europe differs from these that in Asia or North America. In Europe, historical development in hinterland transport was different from in the USA—trucks took the major share from railways, and they still dominate the market (Vassallo 2005). Based on the latest logistics market analyses, they will continue to do so in the future (Stamos 2018). This is even in the situation in the EU where its member countries are trying their best to support functioning and competitive markets for railways, as well as supporting the sector with numerous investments. The ideas and plans as well as EU wide programmes (like Marco Polo) are good, but they have not been enough. Europe functions with road transport, and trucks with semi-trailers. It is interoperable unit all over, and provides door-to-door service in a flexible and prompt manner. During the years, the EU continually enlarged (in other words, can operate across different countries without technical difficulties), and this has meant that lower income countries have joined this economic area. New potential economies have therefore entered the logistics sector, competing with considerable cost advantage against the old West. As the road transport market has virtually been free for market based competition inside of the EU, it has meant cost competitiveness for users, and further dominance of road transport over hinterland (mainly railways, but also inland waterways and pipeline transports).

This chapter presents the effect of sulphur regulation on North European countries (Estonia, Finland and Sweden), which have a clear disadvantage in terms of access to Central European markets (especially in the case of Finland). Data analysis shows that trucks with trailers are increasingly favoured in supply chains, even if it is not the intention or objective of further environmental maritime regulation. First, the situation in Estonia is analyzed, followed by Sweden and, lastly, Finland. The analysis in this chapter highlights that some locations and sea ports have been clear winners, while some others have suffered due to new disadvantages that have appeared as strict environmental measures have been implemented.

4.2   ESTONIAN LOGISTICS SECTOR AND THE GROWING ROLE OF TRUCKS WITH SEMI-TRAILERS

Overall, Estonian sea port volumes have remained unchanged in the period of 1999–2017, if tons handled is used as a measure (Fig. 4.1). In the year 1999, sea port handling was 34.36 million tons and in 2017 it was at 34.8 million tons (an increase of 1.3%). Of course, volumes during the 19-year observation period had clear growth (until 2006), and thereafter had a downswing with some recovery in 2009–2011. What is particularly characteristic of the Estonian sea port-based logistics system is the importance of transit flows, and the significance of one sea port, Tallinn (its operations are not only in the capital city of Tallinn, but also in Paldiski). The transit share from overall handling during the years 1999–2006 varied around 70%, and was highest in 2006 at 77.9%. The same situation applies to the share of the port of Tallinn of overall handling volumes of this small country; it was around 80% during these growth years. However, this dominance of foreign (albeit Russian) transit and one sea port has, since 2006, gradually weakened. In the last observation year, transit share from overall sea port handling was

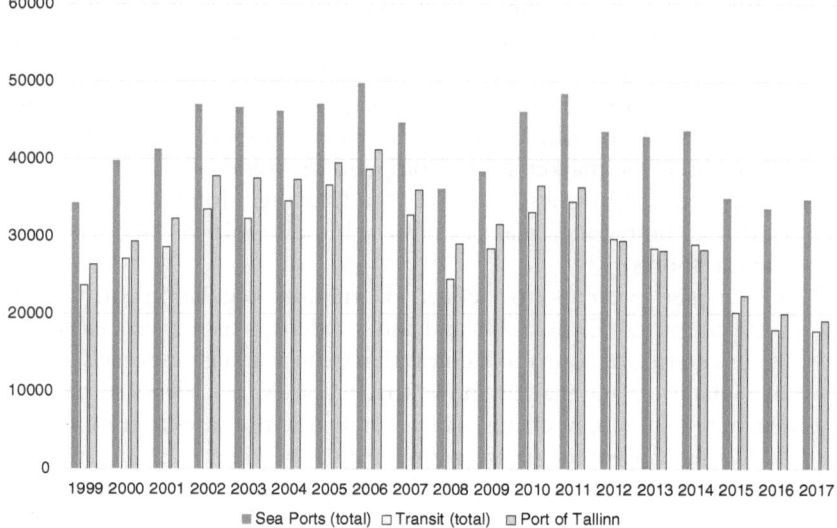

Fig. 4.1 Sea port handling volumes in all sea ports of Estonia, amount of transit and handling amount of port of Tallinn (thousand tons) (*Source (data)* Statistics Estonia 2018; Port of Tallinn 2018)

somewhere above 50%, and the market share of the port of Tallinn was around 55%. Sea ports such as Sillamäe, Pärnu and Kunda have taken their share.

It should be noted that transit volumes as well as port of Tallinn handling volumes declined substantially during the years 1999–2017. Actually, both of these are close to 25% lower in 2017 compared with the base year. Having said this, it does not mean that current operations would suffer in terms of revenue or profits. The logistics system in Estonia has only transformed one step further to serve higher value-added segments, and unitized cargo. Figure 4.2 illustrates the development of revenue and profit (before taxes) at the sea port of Tallinn—in the observation period of 1999–2017, revenues increased by 131.7% approximately to 121 million EUR, while profit increased in the same period by 63% to 38.4 million EUR. These results have not come without substantial investments as total assets of sea ports have grown 155.2% during the years 1999–2017, and were 597.1 million EUR at

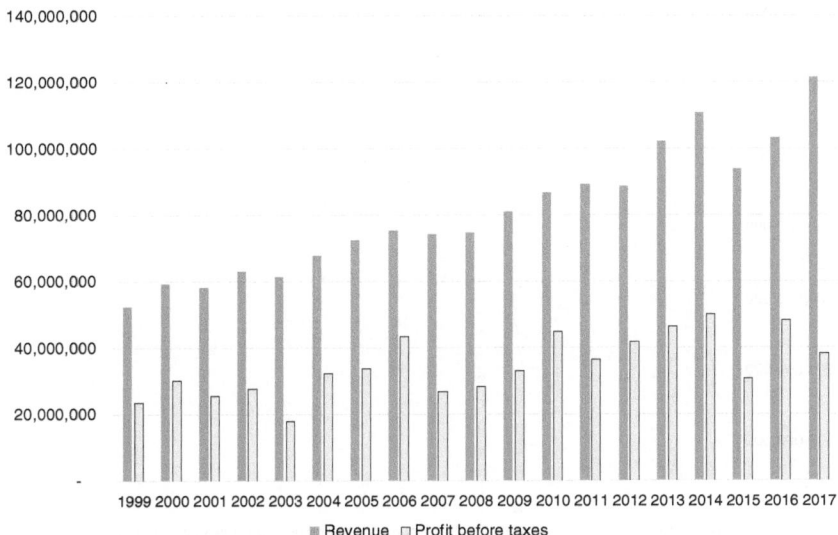

Fig. 4.2 Revenue and profit before taxes at Tallinn sea port (Estonia) from 1999 to 2017 (in EUR) (*Source (data)* Port of Tallinn 2018)

the end of this period. Serving increasing amounts of containers, trucks with semi-trailers, and passengers requires specific investments (Fig. 4.3).

The growing development of container handling and trucks with semi-trailers (RoRo/RoPax cargo) could be detected from Fig. 4.4. A starting level of container handling was very low in the year 1997, its growth rate has been significant, and in 2017, it was nearly 322.1% higher than in the base year (corresponding to growth of 7.5% p.a.). Growth is not only due to the domestic market, or Baltic market containerization, but also increasingly as a consequence of serving transit destinations, and particularly factories on Russian soil (Hilmola and Henttu 2015). During years of tighter sulphur regulation (2015–2016), container handling volumes experienced a drop of somewhere above 20%. This was mostly caused by the Russian market and direction difficulties, but also domestic markets were going through challenging times (as European credit crunch continued in Southern Europe). Some part of this decline could be due to sulphur regulation, and increased freight rates, but there were numerous other reasons (Hilmola and Tolli 2015). Regarding trucks and

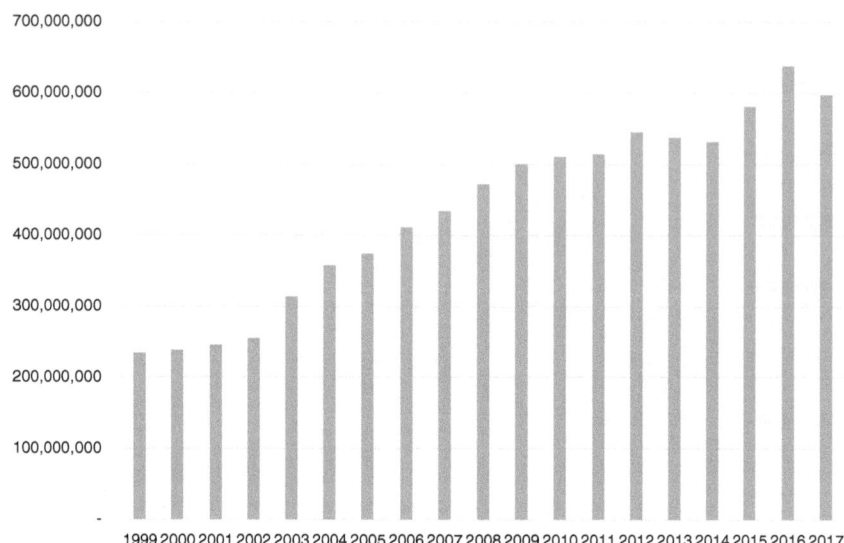

Fig. 4.3 Total assets of Tallinn sea port (Estonia) from 1999 to 2017 (in EUR) (*Source (data)* Port of Tallinn 2018)

semi-trailers, direction and growth has been clear, with only some minor declines during 2009, 2011 and 2015. The tighter sulphur regulation era and difficulties in 2015 in this cargo group could be argued to have belonged to the decline of other RoRo/RoPax directions than Finland (such as Sweden). In fact, the Finnish direction grew well in 2015. In addition, within the port of Tallinn (mostly serving the Finnish route), growth rate in RoRo cargo was 8% in 2015 (Port of Tallinn 2018). In all sea ports, growth trajectory was reached back in 2016 (Fig. 4.4), when the annual change from previous year was 14.2%. In 2017, this fast growth continued as RoRo volumes in all Estonian sea ports expanded substantially, by 22.2%. Most of this growth has been from international supply chains. Growth from year 2005 has been substantial, at approximately 103.6%. In annual terms, truck handling growth has been 6.1%.

If development is examined in these two unitized product groups after 2008, it is evident just by examining Fig. 4.4 that containers have been experiencing the start of a maturing stage, where, in turn, trucks have continued their growth. From 2008, container handling grew by 26.6% (2.7% p.a.), while truck handling grew 64.3% (5.7% p.a.).

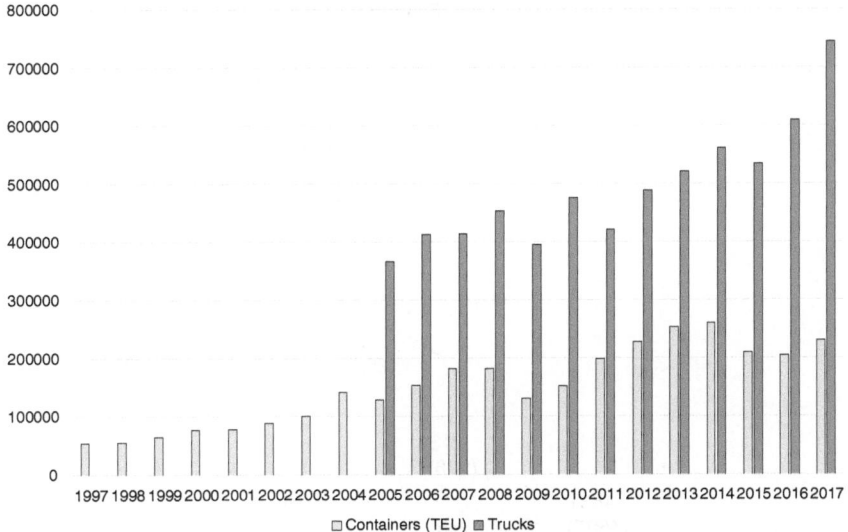

Fig. 4.4 Unitized cargo handling in Estonian sea ports from 1997 to 2017 (trucks have limited data availability, only from 2005) (*Source (data)* Statistics Estonia 2018)

Some of the growth from unitized cargo handling has been devoted to railways, like in containers, but then the direction has been east, Russia (Hilmola and Henttu 2015). However, truck-based RoRo/RoPax volumes have meant that roads are just more used, and road in north–south axis (Via Baltica). This could be justified with railway data from which Fig. 4.5 was compiled. From Estonia in the southern direction, only 92,000 tons were transported on rails (year 2016), and all of this was from Estonia to Latvia. In the other direction, there were 912,000 tons of freight from Latvia to Estonia. However, most of this in total approximately one million tons of volume is something other than unitized cargo—mostly different raw materials (coal and lignite, coke and petroleum products and different chemicals). It is important to note that volumes at railways between Latvia and Lithuania are much higher. Overall, the railway use in Baltic States and within the north–south axis has been on decline for years, and waiting for investments, for example from Rail Baltica, in order to make railway transport shorter in distance (the current Soviet-era network was built to serve national needs and with a focus on eastern directions, not European economic growth), quicker and higher quality.

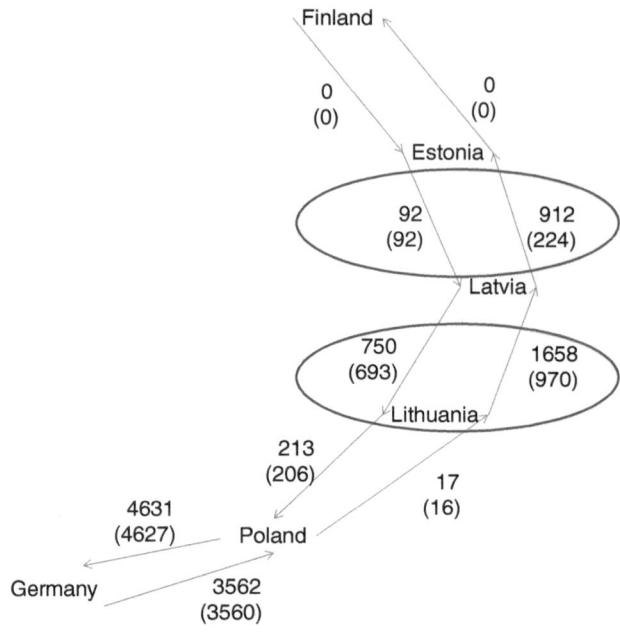

Fig. 4.5 Railway freight volume (in thousand tons) in year 2016 between Finland, three Baltic States, Poland and Germany (in parenthesis is the freight volume between neighbours) (*Source (data)* European Union 2018)

4.3   Swedish Logistics Sector and the Growing Role of Trucks with Semi-Trailers

From the long-term perspective, the growth of sea port cargo handling has been very low in Sweden, as shown in Fig. 4.6. From the year 2000 in total and in all sea ports, material handling has increased 12%, which corresponds to a 0.71% annual growth rate. After the global economic crash of 2009, cargo-handling development has been moving sideways. A similar low growth, in tonnage terms, has occurred in the sea port of Trelleborg since from the year 2000, growth has been 6%. Overall, this is very typical development in Europe, even for a country such as Sweden, which recovered very well from the 2009 crash (in fact GDP growth and employment have been among the best in Europe).

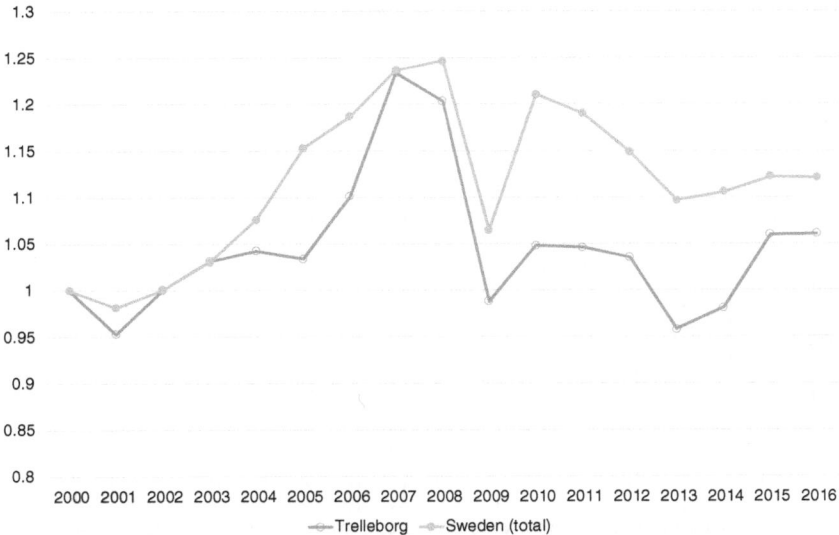

Fig. 4.6 Total amount of cargo handled in Swedish sea ports, and that of Trelleborg during 2000–2016 (indexed, where year 2000 = 1.000) (*Source (data)* Swedish Confederation of Transport Enterprises 2018)

Even if the general growth rate of sea port handling in Sweden has been low, container handling is showing long-term strength (Fig. 4.7). From year 2000 to the end of observation period (2016), growth was 45.3% (2.4% p.a.). The growth rate was of course steepest in years 2000–2007, when container handling grew by 34% (4.3% p.a.). This growth has slowed down in years 2008–2016 to 8.9% (1.1% p.a.). Especially troublesome for growth in this latter period was of course the dip of 2009 (rather conservative, −6.3%), but also no growth in the time period 2012–2015. Actually, in these latter years container handling rate declined slightly. Container handling also declined in 2015 from the previous year (−1.7%); however, it recovered nicely in 2016 (from previous year by +4.7%). The Swedish container handling system is rather concentrated, where the port of Göteborg has been the clear dominating sea port (>50% from the volumes). During the time period of 2000–2016, Göteborg's container handling growth was rather marginal, and it has mostly going on sideways after 2006 (with some peaks and declines during the years; Bergqvist and Cullinane 2017).

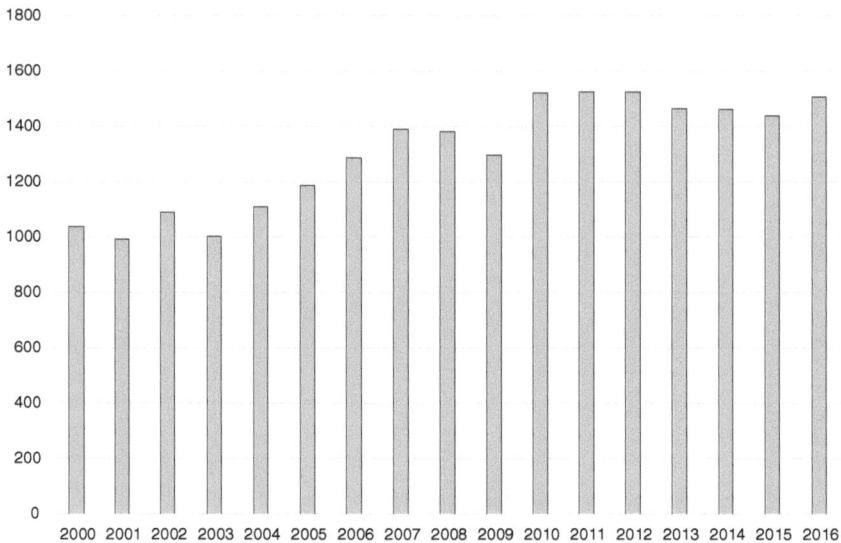

Fig. 4.7 Total volume of containers (in thousand TEU) handled in Swedish sea ports during period of 2000–2016 (*Source (data)* Swedish Confederation of Transport Enterprises 2018)

Earlier research indicates that there are reasons for this relating to sea port privatization (concession contracts) and labour unions (Bergqvist and Cullinane 2017). Therefore, most of the growth in the Swedish case has come from other sea ports, like Helsingborg and Gävle. One particularity of the Swedish container handling system is that it is concentrated on west coast, which has better and shorter-distance access to the deep sea.

In contrast to the general very low growth rate, handling of trucks (most often accompanied with semi-trailers) has increased in the same period (since 2000) in Sweden by 35.7% (1.9% p.a. growth). This is shown in Fig. 4.8. The leading RoRo cargo sea port in Sweden is Trelleborg, which is located in the far south and has close proximity by sea to Germany and Poland. As can be seen from Fig. 4.8, its volumes have grown in the period of 2000–2016 by 57.7% (2.89% p.a. growth), and it accounts for around 24–25% of all RoRo volume in Sweden. Apart from 2008 to 2009 and the port of Helsingborg, growth has been rather consistent in all sea ports of Fig. 4.8 (especially in 2010–2016). Implementation of sulphur regulation in 2015 (of 0.1%) was really the

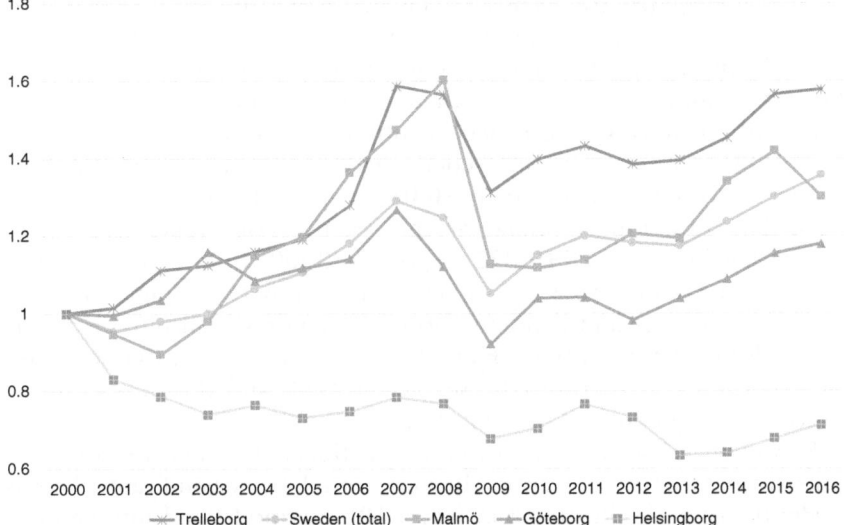

Fig. 4.8 Amount of trucks handled in four Swedish sea ports and handling overall from 2000 to 2016 (indexed, where year 2000 = 1.000) (*Source (data)* Swedish Confederation of Transport Enterprises 2018)

year of growth for all of these sea ports since on average volumes grew by 5.4%, and the best growth was at Trelleborg, at 8%. Earlier research has found some reasons for the long-term success of Trelleborg, including the high frequency of RoRo shipping routes, which do not take that long in time, and sea port as well as shipping schedules fit well on the needed breaks of truck drivers (Woxenius 2012).

It is understandable that not all sea ports have been successful in RoRo cargo. As shown in Fig. 4.8, the sea port of Helsingborg has been in declining truck handling development after the year 2000. This could be explained in part by the opening of Oresund Bridge (and tunnel) between Sweden and Denmark. It hurt cargo handling in the very short distance sea port of Helsinginborg since competition enabled a direct hinterland connection to Denmark. The role of Denmark as a transit country to Swedish cargo has been challenging since extra distance is needed (or additional ferries taken) in order to reach Central Europe. Other sea ports in Fig. 4.8 offer direct RoRo connections, with Trelleborg being the shortest sea route to the main markets.

Using railships to transport railway consignments directly to destinations was really popular from Sweden—as well as from Finland—in the 1990s. This is not the case anymore. As Fig. 4.9 illustrates, annual number of railway wagons loaded or unloaded to ships with their cargo have been continuously declining. During the year 2000 in all sea ports of Sweden, this activity was 221,012 wagons and in 2016 it had dropped to 30,037 wagons (−86.4%, or −11.7% p.a.). The situation is similar at the sea port of Trelleborg, where decline has been from 119,022 wagons in 2000 to 20,871 wagons in 2016 (−82.5%, or −10.3% p.a.). The reason in here is simple: it is too costly to transport and handle wagons at sea ports. Railships also consume a lot of diesel oil (e.g., Hilmola 2012), since they carry structure on structure on structure (the sea vessel itself, and the heavy railway wagon carrying container or semi-trailer).

However, it does not mean that international and intermodal railway units from Sweden to Central Europe would be insignificant. This is hardly the case. As illustrated in Fig. 4.10, intermodal and international

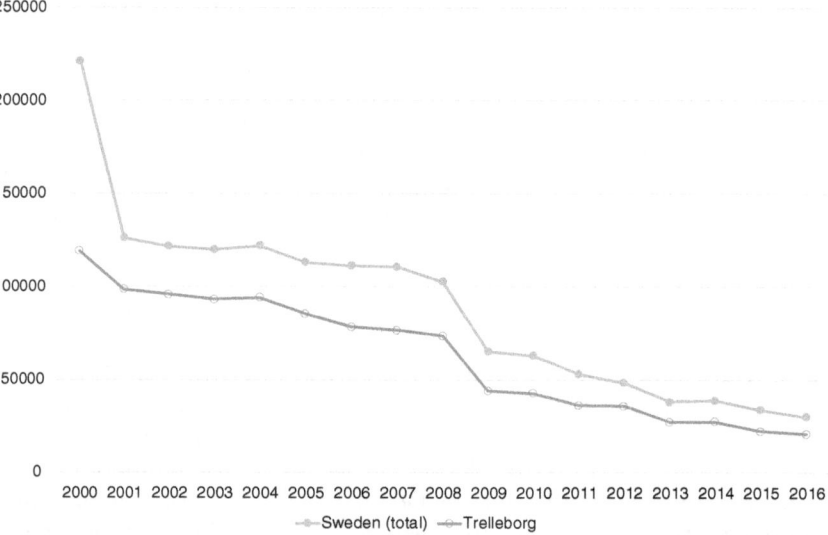

Fig. 4.9 Railway wagons handled and loaded/unloaded to the ships in total in Sweden and in the port of Trelleborg from 2000 to 2016 (*Source (data)* Swedish Confederation of Transport Enterprises 2018)

transport at railways have been the success-story of Sweden in recent decades. Growth was especially brisk in 2005–2010. This cargo is transported to, for example, near Oslo, Norway, or alternatively by Oresund Bridge to Denmark, and then further down to Germany and/or Central Europe. It could be estimated that current level of 4–5 million tons of international intermodal cargo is around 300,000 semi-trailer trucks carrying forty-foot equivalent unit containers.

Railway standards, such as gauge width, is the same across Sweden, Central Europe and Norway (1435 mm), while in Finland it is the old Russian standard gauge width (1524 mm). Even if Finland and Sweden do have hinterland connection in the north for railways, volumes are currently very low since additional transshipments are needed and they increase the costs. Figure 4.11 presents development in railway freight volumes between Finland Sweden. Sulphur regulation did not bring a resurgence to the development, and volumes in the years 2015–2016 were very low (basically, no longer existent).

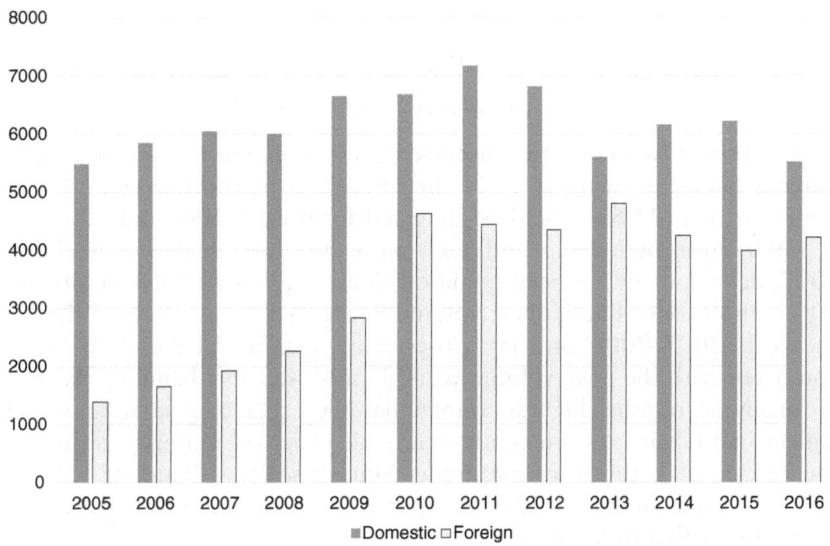

Fig. 4.10 Intermodal transports at Swedish railways during the years 2005–2016 (in thousand tons) (*Source (data)* Transport Analysis 2018)

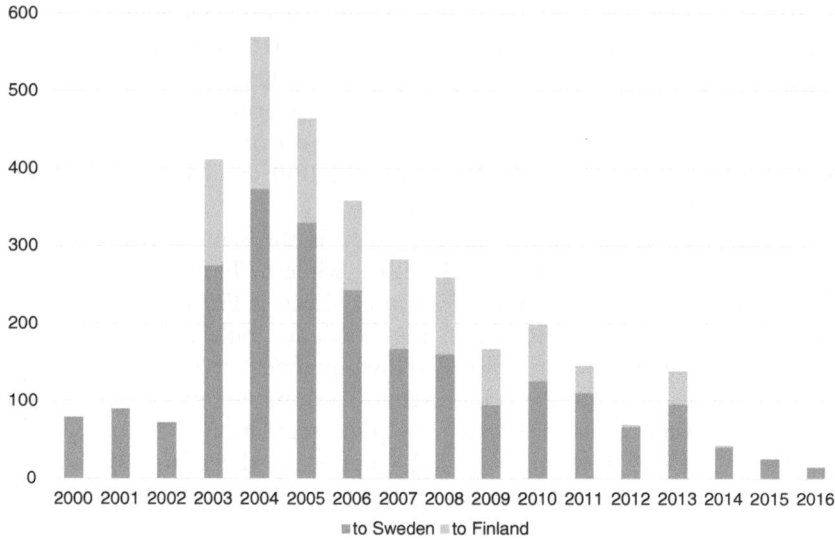

Fig. 4.11 Railway freight transports between Sweden and Finland during years 2000–2016 (thousand tons) (*Source (data)* European Union 2018)

4.4   FINNISH LOGISTICS SECTOR AND THE GROWING ROLE OF TRUCKS WITH SEMI-TRAILERS

Similar to the two countries discussed previously, Finnish sea port handling is hardly growing at all in the long-term perspective (Fig. 4.12). Some growth (22.5%) could be detected from early 2000 until the end of observation period (around 1.2% p.a.); however, in the years 2003–2017, development has been going on sideways, varying between 80 and 100 million tons. The highest volume in Fig. 4.12 was achieved during the years 2007–2008, and last observation year is still 3.6–3.7% lower compared with the peak volume years. Finnish sea port handling has traditionally been rather high as country has been characterized as an island nation in Europe, and currently nearly all of its oil imports are implemented through the maritime mode. Finnish sea port handling volume was 2.82 times higher in 2016 compared with Estonia. However, it was 28.6% lower that of Sweden.

Container handling shows some growth from the early years (2000–2001); however, even this growth could be considered to be

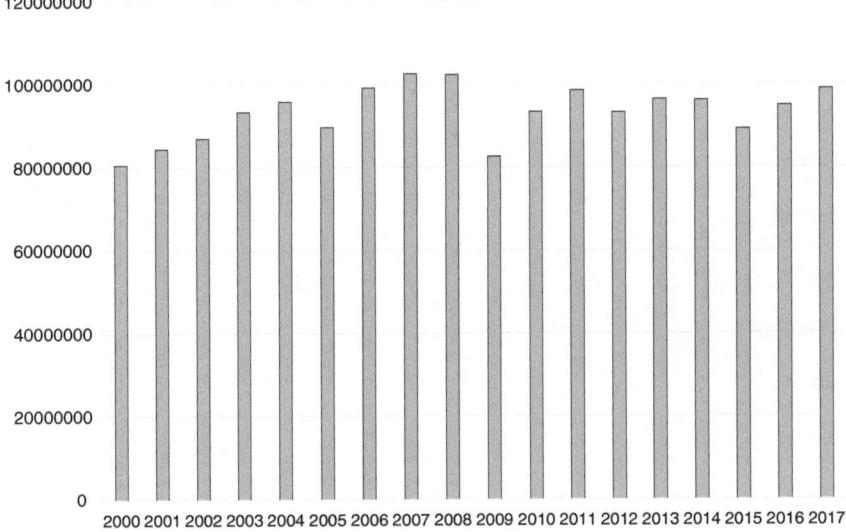

Fig. 4.12   Total volume of cargo handled in Finnish sea ports from 2000 to 2017 (*Source (data)* Statistics Finland 2018)

conservative (3.7% p.a. in the period of 2000–2017; see Fig. 4.13). The growth trajectory was steeper in years 2000–2008, but this was caused by a combination effect of increasing containerization, competitiveness of Finnish export industries, a drive in consumption and imports as well as the favourable role of Finland as the Eastern transit route. Most of these drivers softened in the years 2009–2012; however, containerization continued and trade growth surprised on the upside in the last observation years. Therefore, the data series shows at least some identifiable growth. However, it should be noted that in 2009, but also in the years 2014–2015, container volumes declined, which had not been that common in earlier years (actually the economic problems of IT/dot.com bubble burst in early 2000s had no effect on container handling). The growth story of containerization is still in place, but it is much more vulnerable to macro-economic events.

As far as trucking units are concerned, there are two ways in the Finnish logistics system for how units are handled, and what units are eventually used. If the sea journey is a longer one, such as to Germany, it is common that only semi-trailers are loaded on RoRo/RoPax ships.

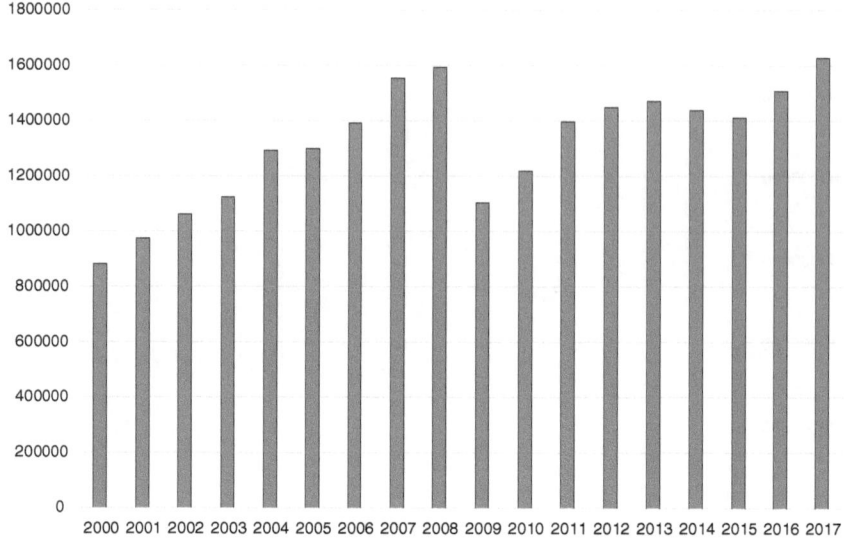

Fig. 4.13   Total volume of containers (TEU) handled in Finnish sea ports from 2000 to 2017 (*Source (data)* Statistics Finland 2018)

Loading is completed in sea ports by special purpose tractors or push/ pull types of units. This approach has been in a difficult position since 2007 as Fig. 4.14 shows. The volume in the last observation year is practically the same as this year. So, the last decade has shown sideways growth, nothing else. Growth exist in data series, but only if observed from early 2000 (sharing similar growth to overall tonnage handling, at 1.6% p.a. since 2000). This unitized cargo type has also shown softness during the years, since in 2009 volumes declined considerably, and to some extent also in years 2008, 2012 and 2013. So, the growth story is in place, but again it is limited to external events (such as the mac-ro-economy, oil price changes, environmental legislation etc.).

Another way of progressing unitized trucking cargo is to use both trucks and semi-trailers all the time in the transportation process. It is of course costly (as both expensive truck and driver are tied up in the transportation process together with semi-trailer), but it is more flexible, saves transshipment costs and is not affected that much by local events at sea ports (such as strikes or limited working hours). This approach is favourable in short-distance sea journeys with RoRo/RoPax ships,

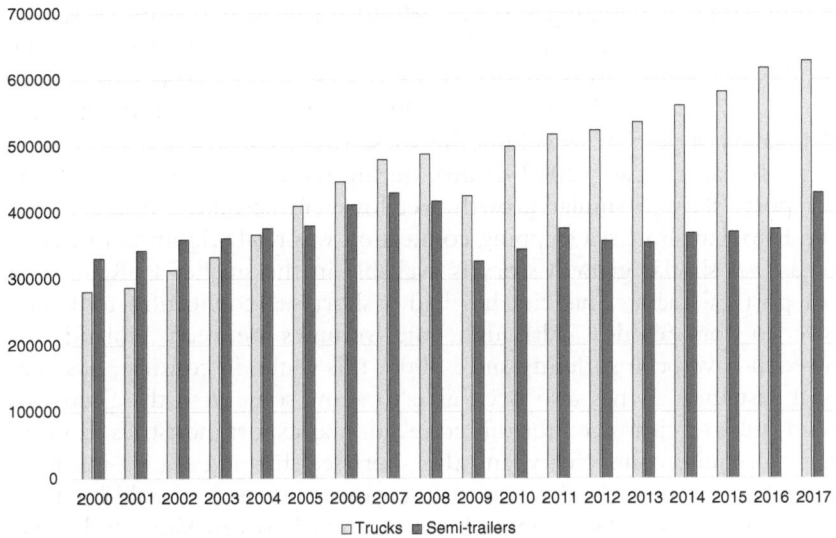

Fig. 4.14 Total volume of trucks with trailers, and semi-trailers handled in Finnish sea ports from 2000 to 2017 (*Source (data)* Statistics Finland 2018)

and especially to countries where truck driver salaries are lower. This approach also showed its robustness during the implementation of sulphur regulation. Volume growth from 2000 onwards speaks for itself, having more than doubled until 2017, and showing annual growth of 4.8%. The only decline in the volumes took place in 2009, when trucks with semi-trailers experienced a 12.8% decline. This was much lower than that experienced by the trailers (−21.5%) or containers (−30.7%) at the same time. Growth is therefore pretty much Consistent, and since 2009, volumes have only demonstrated growth. This is despite all the macro-economic uncertainties of European Union countries, Russia or the implementation of sulphur regulation at sea.

Growing short sea shipping within the range of a maximum of 100 km was not only an opportunity for big sea ports and dominant players, such as Helsinki. Some smaller sea ports (offering linear shipping services for the truck with trailers combinations) also wisely built attractive connections after 2010. Figure 4.15 presents two good examples. The first one is sea port of Hanko, which has been for a long time a vital connecting point to Germany (and this with trailers without trucks as longer linear shipping service). However, in recent years its service portfolio was enlarged with RoRo connection to Estonia, Paldiski. This resulted in the enormous growth of trucks with trailers, if examined from 2007 or from the base year 2000 (both would show growth in the thousands of percent as volumes in these two comparison years were so low). In earlier years (2002–2006), during the boom of Russian transit, sea ports showed similar growth (yet lower in absolute numbers), but the Estonian short sea shipping connection was really significant in every regard. A similar growth story is available in the smaller RoRo/RoPax sea port of Vaasa. This city has had a short sea connection to Umeå, Sweden for decades. Although, the volumes between Finland and Sweden have been suffering since 2007, this route, in contrast, has been well sustained. It has also been able to show some growth. There was (and still is) a clear need for this connection as export industries from the region require connectivity, and this short sea shipping sea leg is able to provide short hinterland access to European markets. It should be noted that in late 2011, the shipping line operating between Vaasa and Umeå

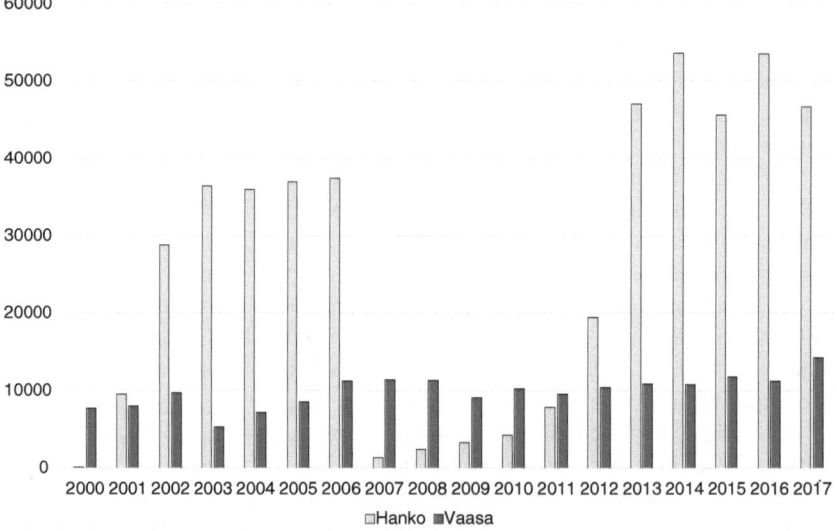

Fig. 4.15 Total volume of trucks with trailers in two selected sea ports (Hanko and Vaasa) of Finland from 2000 to 2017 (*Source (data)* Statistics Finland 2018)

went bankrupt, and was re-established again in 2013 by these two cities, and has been a success so far. This route serves transportation needs down to Italy. Growth between the years 2014–2017 has been 31.3%, which highlights the importance of short sea shipping connectivity in an era of demanding sulphur regulation.

4.5   Concluding Discussion

Freight handling at sea ports hardly shows any significant growth in these three analyzed countries. Development has been mostly leveled off or declining, depending on the country in question. This does not mean that the logistics sector is unhealthy; however, it is greatly changing. Unitized cargo is taking an increasing share, and particularly trucks with trailers driven inside of RoRo/RoPax ships have experienced growth in recent decades. This growth shows no sign of slowing down as environmental demands at sea are increasing. From an environmental perspective it would be much better to utilize more hinterland transportation such as railways, instead of road, however, based on statistics, there is no trend to suggest this yet. In Sweden, volumes of international combined transport at railway network are already significant, but in very recent years, they have not shown that much growth. In addition, railship traffic (loading railway wagons on ships to reach Central Europe) have dropped in volume significantly. In the case of Estonia, the situation is similar as international railway freight volumes have declined further. In fact, in the Estonian case, the starting level in volumes was already low, and getting lower has been a bit of an unwelcome surprise. The story for railway traffic is pretty much the same (without the sea component) between Finland and Sweden.

This chapter discusses the sustainability of current supply chains. They may have a short lead time, and be prompt and flexible, but road transport in general emits a lot, requires spaceand also causes accidents. It is by no means sustainable. Here, it seems that implementing a demanding sulphur regulation has benefitted the sea area (in lower emissions), but on the other hand, it has created another problem in the hinterlands. Ultimately, society will end up paying negative externalities. Countries in Europe have started to gather more road use payments, and hinterland transport seems to be additional cash cow to fill governmental budgets. The development path is not a desired, nor a planned one, but rather has happened as a coincidence.

References

Bergqvist, R., & Cullinane, K. (2017). Port privatization in Sweden: Domestic realism in the face of global hype. *Research in Transportation Business and Management, 22,* 224–231.

European Union. (2018). Eurostat database. *Transport.* Available at http://ec.europa.eu/eurostat/web/transport/data/database. Retrieved 26 Feburary 2018.

Hilmola, O.-P. (2012). *Competing transportation chains in Helsinki-Tallinn route: Multi-dimensional evaluation* (Research Report 243). Lappeenranta, Finland: Lappeenranta University of Technology, Department of Industrial Management.

Hilmola, O.-P., & Henttu, V. (2015). Border-crossing constraints, railways and transit transports in Estonia. *Research in Transportation Business & Management, 14,* 72–79.

Hilmola, O.-P., & Tolli, A. (2015). Early 2015 performance in Baltic Sea ports: Forecasts of Estonian performance for entire year. *Journal of Transport and Telecommunication, 16*(3), 183–189.

Port of Tallinn. (2018). *Annual reports 1999–2016.* Available at http://www.portoftallinn.com/annual-reports. Retrieved 27 Feburary 2018.

Stamos, I. (2018). What do data tell us? The story of the European logistics and road freight transportation sector. *Proceedings of 7th Transport Research Arena TRA 2018,* April 16–19, 2018, Vienna, Austria.

Statistics Estonia. (2018). *Statistical database: Economy.* Available at http://pub.stat.ee/px-web.2001/I_Databas/Economy/databasetree.asp. Retrieved 27 Feburary 2018.

Statistics Finland. (2018). *Foreign shipping traffic (Finnish Transport Agency).* Available at http://pxnet2.stat.fi/PXWeb/pxweb/en/StatFin/StatFin__lii__uvliik/?rxid=e5a1874c-881a-4ad3-b2d9-c652a2d6e813. Retrieved 28 Feburary 2018.

Swedish Confederation of Transport Enterprises. (2018). *Sea port statistics of ports of Sweden.* Available at https://www.transportforetagen.se/In-English/Association-ports-of-Sweden/Statistics/. Retrieved 22 Feburary 2018.

Transport Analysis. (2018). *Rail traffic in Sweden—Official statistics of Sweden.* Available at https://www.trafa.se/en/rail-traffic/rail-traffic/. Retrieved 22 Feburary 2018.

Vassallo, J. M. (2005). Nature or nurture: Why do railroads carry greater freight share in the United States than in Europe? (Research Working Paper Series, WP05-15). Harvard University, USA.

Woxenius, J. (2012). Flexibility vs. specialisation in Ro-Ro shipping in the South Baltic Sea. *Transport, 27*(3), 250–262.

CHAPTER 5

Maritime Supply Chains: How They Experienced the Regulation Change

Abstract The change in regulation in the Baltic Sea concerning sulphur emissions from shipping was a demanding change. Finnish and Estonian companies took a 'wait and see' strategy approach, where only one major shipping company invested in scrubbers on a large-scale. Many shipping companies have made initial investments in Liquefied Natural Gas (LNG) ships, which are serving, or are going to serve in the near future, RoPax, container and bulk customers. Despite these investments, in the previous decade, shipping company asset amounts have been declining, and revenue growth has been minimal. Yet some actors have shown abnormally high profits, even in this very demanding environment. Hinterland transportation logistics actors did not benefit from sulphur regulation, and an analysis shows that their profit margins are thin and their business risks are high. Some terminals receiving and distributing LNG ships were built during the build up to the environmental emission reduction, but this was for a number of reasons. It remains to be seen whether the transition to LNG is economically sustainable, because its prices fluctuate, and show the same uncertainty as can be seen with oil. The private sector expects a lot from LNG, and this together with seeking new cargo routes were seen as the most important strategies tackling change.

Keywords Shipping companies · Hinterland logistics · Sulphur regulation · Baltic Sea · LNG

© The Author(s) 2019
O.-P. Hilmola, *The Sulphur Cap in Maritime Supply Chains*,
https://doi.org/10.1007/978-3-319-98545-9_5

5.1   Introduction

Since maritime transport has become so critical for global economic growth, its environmental emissions were not previously the subject of much attention earlier in policy and legislation changes. Unit sizes in shipping have continued to increase, and transport has already become cheaper and with lower emissions by itself. In addition, the total emissions of shipping from world CO_2 emissions are marginal, and for this reason, for a long-time shipping was not the subject of any attention among policy makers. However, there were many negative issues forming, despite the low overall emissions. Since the shipping network is heavily concentrated (as is all logistics) to a few nodes, it means that these low-sounding emissions are mostly concentrated to some small areas. This causes health issues for the population living in cities nearby, and, as further concentration and larger ships became a reality, these only increased their negative externalities. Shipping used very low quality heavy fuel oil for a long time, and its sulphur content was as high as 4.5%. Before the year 2020, the world is using a maximum 3.5% sulphur fuel, apart from in Emission Control Areas (ECAs, coasts of North America as well as the North and Baltic Sea), where after 2015 a maximum 0.1% sulphur content measure was implemented. The world will change in 2020 since globally regulation comes into force that sets a maximum 0.5% sulphur content level, which basically means that the experiences of ECAs are in need around the world to avoid excessive cost level increases in maritime supply chains.

Previously, widely used heavy fuel oil comprised oil residue from refineries, and it was (and still is) cheaper than regular oil, like Brent. This of course helped shipping to keep its dominance in global logistics since fuel cost was not an issue, and shipping speeds were rather low cost and could be increased as needed (strongly growing economies or just need for speed). This old world is coming to an end, and from the year 2020 onwards, the entire world needs to rethink its maritime supply chain strategies from a new angle. If cheap dirty heavy fuel oil is desired, then all ships need to invest millions of USDs installing scrubber equipment onboard each ship (Fig. 5.1). This big cleaning machine simply uses sea water to clean sulphur from exhaust gases (water could be returned to sea or alternatively scrubber could be the closed loop type, with the same water used time and time again). Sulphur is then recycled at sea port operations to waste treatment stations. This new equipment

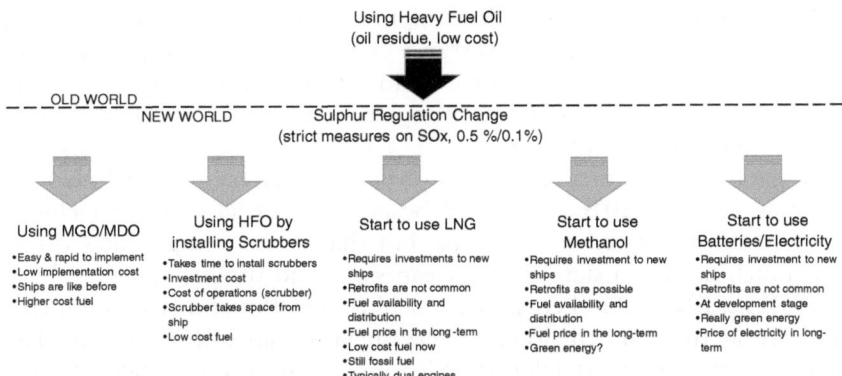

Fig. 5.1 Alternatives to old world thinking in the face of sulphur regulation

needs electricity to operate, and it requires maintenance. In addition, the fleet where the scrubber is invested cannot be very old because its money-saving lifespan would be too limited. The scrubber will also take some space from the ship (mostly taking away from revenue-generating cargo space). Taken together with these limitations and constraints, the benefits of low cost oil could be continued with scrubbers (Ma et al. 2012; Yang et al. 2012; Lindstad et al. 2017).

Companies may find it easier just to change to using cleaner Marine Gas Oil (MGO), which has the necessary low sulphur content (Zis et al. 2015). The only downside here is its higher cost. The difference to earlier fuel cost is significant. This change could be softened with fuel-saving approaches in operations (e.g., lower speeds and different shorter routes) and by applying optimization in the shipping network operations. In addition, diesel price variations could be tried to be utilized by acquiring fuel where it is cheapest, and when it has a low cost at world markets. Customer contracts need urgent attention, and additional monthly/weekly surcharges on oil need to be implemented without any delay. It should be emphasized that a diesel oil grade switch is in the mind of and within the action list of most of the shipping companies as they prepare for the coming change of 2020 (Terazono and Hume 2017; Churchill 2017). However, there are some reasons to assume that not all actors in the shipping world will follow these strict cost-increasing measures; cheating with oil grades is possibility, especially in deep sea voyages (Churchill 2017).

Other alternatives to tackle low sulphur challenge are more long-term oriented. One of them is to invest in a completely new fleet, using some other fuel type (Fig. 5.1; the three options from middle to right). At the moment at the Baltic Sea, it is very popular to invest in Liquefied Natural Gas (LNG) ships (Gritsenko 2018). However, other alternatives also exist, such as using methanol ships or entirely battery powered electricity traction (Line 2016). It would be possible for all of these technologies to be retrofitted to old ships, but that is not being widely done. It is much better to start from a clean slate, and produce purpose-built ships. Both LNG (Burel et al. 2013; Schinas and Butler 2016) and methanol are at the moment looking very lucrative since they produce less CO_2, are able to avoid sulphur emissions and have very small nitrogen emissions. However, other harmful emissions could be much higher as research reports high methane emissions while using LNG-powered ship (Anderson et al. 2015). An electricity-powered ship would avoid all of these emissions, if energy came renewable sources. From a technological perspective, LNG and methanol are already available, and ready for application and use. However, their drawback is the requirements for a fuel distribution network infrastructure, which is expensive to build, and takes long time to implement (Burel et al. 2013). It is also questionable as to whether low prices of these fuels would remain, if their use became more mainstream. Production site cost of LNG is just one part, and another significant addition is the one to supply it to some very northern sea port. If LNG and methanol are evaluated from the environmental perspective, they are both ultimately fossil fuels (methanol is often produced from LNG). However, methanol could be produced from biosources, which makes it a bit better.

This chapter is structured as follows: in Sect. 5.2, shipping companies are reviewed through financial measures as they experienced low sulphur environment change. Most of the companies are from Finland and two are from other countries (Estonia and Netherlands). This is followed in Sect. 5.3, where two publicly traded hinterland transportation companies from Finland are analyzed with the same financial measures. The empirical material grows with the analysis of some qualitative interviews conducted during the years (in Finland and Estonia) following implementation of sulphur regulation in Sect. 5.4. As it has become increasingly popular of using LNG in maritime supply chains and it is getting so much attention in companies, it is reviewed in Sect. 5.5 from the sea port perspective (LNG terminal operating in southern Baltic States), while longitudinal prices of gas are analyzed in Sect. 5.6. The prices of gas around the world are considered, and they are also compared with oil prices in the same period. Section 5.7 finalizes this chapter with conclusions.

5.2   SHORT SEA SHIPPING COMPANIES

In this section shipping companies, mostly operating in the Baltic and North Sea area, are analysed. Nearly all (six out of seven) have headquarters in Finland or Estonia, and vary in size. Some of them are larger corporations, where shipping is just one part of the business (nevertheless, the most significant one, which enables others), and other branches could be hotels, restaurants, retail, warehousing, hinterland transports, and taxi transport for passengers. Some of the companies have also enlarged their service portfolio outside of the Baltic and North Sea to such places as the Mediterranean. The size of these companies vary as some have a revenue of below 100 million EUR revenue (the lowest is around 10 million EUR), and largest one has nearly one billion EUR. The three largest companies in the following analysis take more than 8% from revenue (out of seven shipping companies).

In revenue terms, these analysed companies experienced growth era during the years 2005–2011 (Fig. 5.2). Revenue increased from 1.26 billion EUR by 67.7% in this time period up to nearly 2.5 billion EUR (one reason for the 2011 spike in total revenue is the accounting period of the largest company in the dataset, which had an accounting period 16 months in 2011 and it had started already in 2010). Thereafter, total revenues of these seven companies have somewhat declined and remained in the level of around 2.2–2.25 billion EUR. The reasons for no growth environment are not only related to environmental regulations, but also to the European macro-economy, the economic development of these North European countries, and the dispute in Ukraine as well as the following sanctions that have occurred as a result of this situation. At the company level, RoRo II has lost most of the business revenue in the period; however, this could be related to its corporate structure and the fact that its business was sold to Central Europe in early 2000 (so revenues in years other than 2005 do not include actual shipping, but stevedoring and sea port-related activities). The bulk shipping company has basically the same revenue in 2017 as it had in 2005. The other five companies have shown growth in years 2005–2017. However, growth rates after the implementation of demanding

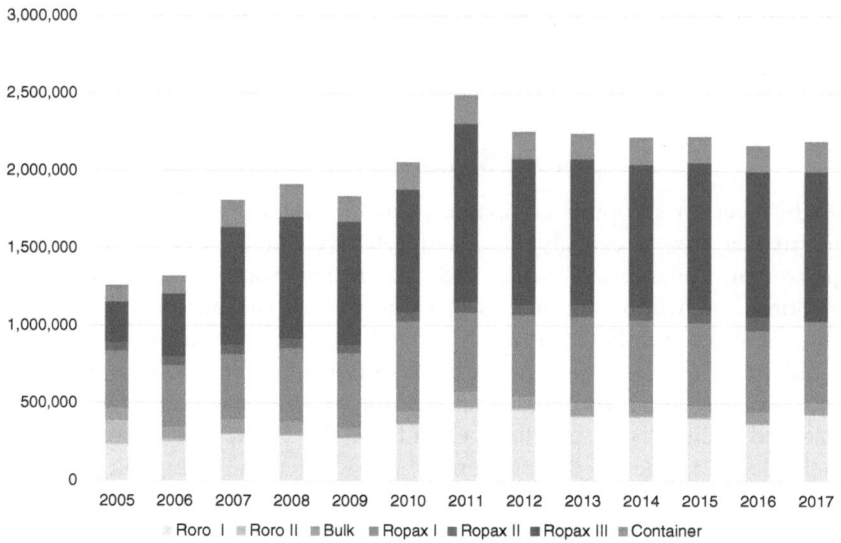

Fig. 5.2 Annual revenues of most active shipping companies in Northern Baltic Sea (thousand EUR), serving three countries of interest during years 2005–2017 (year 2017 data: RoRo shipper II and RoPax II not available) (*Source (data)* Annual reports, profit and loss statements)

sulphur regulations (in 2015) have been low or even negative. The best growth in this period (revenue in 2017 vs. 2014) took place in the company 'Container' (+8.7%), followed by 'RoPax III' (+4.9%) and 'RoPax II' (+4.7%) as well as 'RoRo I' (+4.2%). Most of the revenue was lost at 'RoRo II' (−22.7%, but this is with year 2016 data), and 'Bulk' (−6.9%).

Typically, transportation and logistics sector companies produce the best profits when there is a demand surge in markets (such as in shipping due to military conflicts starting or ending) and/or oil becomes really cheap (it is able to find most of latent transportation demand, but also gives high profits due to better margins). Overall, the best year for profits was 2006, when oil was still reasonably priced (before the surge of 2008), and demand in the transportation markets was strong (Fig. 5.3). The total profit back then was 173.8 million EUR. Only in very recent years, at the time of implementation of demanding sulphur regulation and the nea-simultaneously oil price drop, the profits of some actors increased significantly. For example, companies 'RoRo I' and 'Bulk' produced the highest

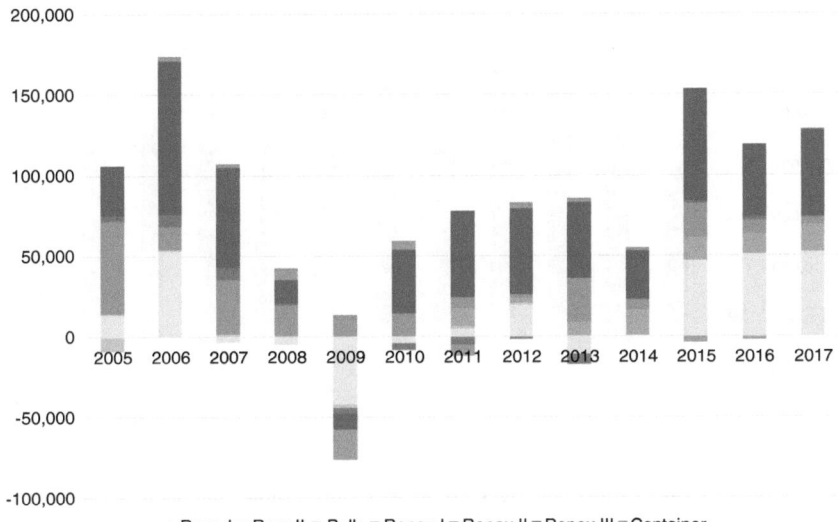

Fig. 5.3 Profit and loss of most active shipping companies in Northern Baltic Sea (thousand EUR), serving three countries of interest during years 2005–2017 (year 2017 data: RoRo shipper II and RoPax II not available) (*Source (data)* Annual reports, profit and loss statements)

absolute profits in the entire observation period during year 2017. In case of 'RoRo I,' these profits were significantly higher than in any year during the period of 2005–2014. Company 'Bulk' has shown much better profit levels in the years 2011–2017, and the situation is quite similar to that of 'RoRo I.' Development has not been the same for all of the actors. Some companies are going through challenging times regarding profitability: the years 2015–2016 were not that good for companies 'Container' and 'RoRo II.'

Regarding total assets, these seven companies hold quite an extensive amount of assets, worth 3.4 billion EUR in the last years of the observation period (Fig. 5.4). However, these are concentrated to a few actors. Companies 'RoPax III' and 'RoRo I' take nearly 78% from the overall value of assets among the seven companies. What is notable in the asset development is the clear growth in the period of 2005–2009 as they increased in a short amount of time to nearly 4 billion EUR and the growth recorded was 138.2%. After this, total assets have declined

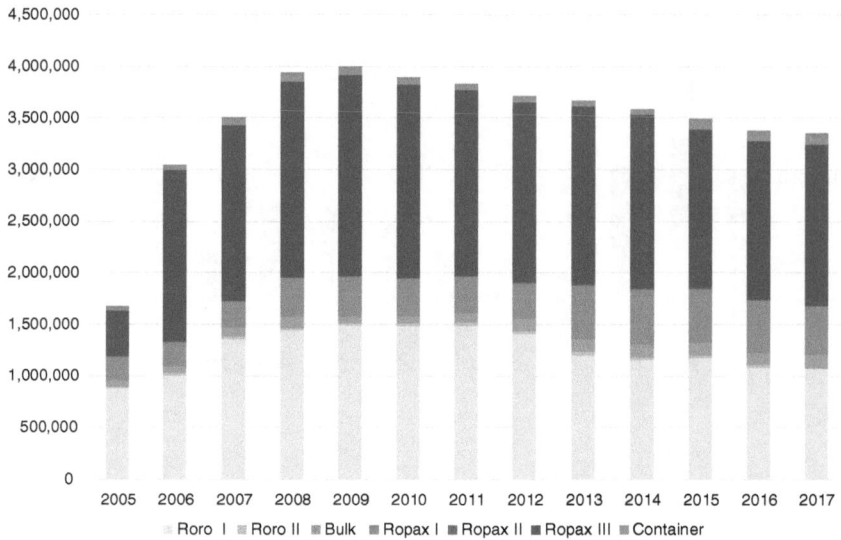

Fig. 5.4 Total assets of the most active shipping companies in the Northern Baltic Sea (thousand EUR), serving three countries of interest during years 2005–2017 (year 2017 data: RoRo shipper II and RoPax II not available) (*Source (data)* Annual reports, profit and loss statements)

continuously, and in 2017 they were around 12% lower than the peak of 2009. Many companies have restructured their operations by selling some assets (ships) in this period, but also new building market deliveries have been few (but having at least some new ships).

Reading through and examining the annual reports (which are public) of some of these companies (as they are available; only RoRo II does not have such records), it could be found that these companies reacted on to the challenge of the 2015 sulphur regulation with different strategies. The strategy of installing scrubbers to existing ships was applied mostly by 'RoRo I.' This was carried out based on a plan with quite a substantial budget, and happened gradually in the years 2015–2017. During the installation process, ships used MGO oil, and some ships in the fleet still do (if scrubbers were not installed eventually on them). Company 'RoRo I' also reveals, that they are cleaning and treating the bottoms of ship with silicon (to avoid excessive fraction, and eventually diesel oil use in operations). They have also installed new propulsion systems to ships.

All RoPax companies ('RoPax I–III'), and 'Bulk' have mostly utilized the strategy of using MGO diesel, instead of installing scrubbers (similar finding in Olaniyi 2017). Company 'RoPax I' also reports to have regularly washed ship bottoms to save friction and energy. They also report using the tactic in sea ports that diesel motors are turned off from the ship after successful docking, and the entire vessel is connected to the electricity grid in order to get clean power. Company 'Container' has at least one scrubber being installed. Apart from these bigger companies that were analyzed, there were of course some smaller shipping companies in Finland. One of these smaller companies installed scrubbers to all of its five vessels. These ships all are container vessels.

What is interesting for the future is the small LNG boom that has been experienced among these companies. Companies 'RoPax I' (delivery already in year 2013) and 'RoPax III' (delivery in year 2017) have both already invested in modern LNG ships, and these RoPax ships are operating in the routes of Turku–Stockholm (RoPax I) and Helsinki–Tallinn (RoPax III). Both of these ships were constructed at a shipyard within Finland. The prices of these vessels were rather high, in the range of 230–240 million EUR (based on earlier research Schinas and Butler (2016), LNG ships are evaluated to be 15–30% more expensive compared with traditional diesel). Company 'RoPax I' is continuing its LNG investments, and it has ordered another LNG vessel from China to be delivered for operations in the year 2020. The price of this delivery is somewhere below 200 million EUR. Companies 'Bulk' and 'Container' are also building future maritime supply chains on the basis of LNG. Both are constructing new ships in Chinese shipyards, where 'Bulk' has ordered two new bulk ships, and 'Container' four new container vessels. Investment amounts in these are not as high since supplied vessels are not for passenger transports—the total investment of 'Container' is reported to be 150 million EUR. Some of these ordered ships will enter operations as soon as 2018. There are clear advantages in LNG-powered ships—one of them is the much lower CO_2 emission level as compared with diesel-powered ships. Typically, these LNG ships do not emit sulphur at all, and emissions of nitrogen are at low levels. In addition, operating these ships brings cost savings in terms of fuel costs. Some companies are planning to enlarge LNG use to hinterlands too; one such company is 'Container,' and it is actually planning to implement the first entire LNG-based supply chain. They already have some LNG-fueled trucks in operation.

It is very difficult to say what the best long-term strategy is. However, from a short-term strategy perspective, it could be said that company 'RoRo I' has been showing the best profitability growth while also experiencing significant improvement in absolute profits during the years 2015–2017. This is clearly apparent in Fig. 5.3. Its strategy is also rather different from the others in the way that it has announced it will take some of the existing ships back to shipyards to enlarge them in order to make space more for cargo. These modifications and renovations are typically significantly much cheaper than building an entirely new fleet. It will also bring some new capacity to the markets. 'RoRo I' has also recently announced some new built contracts for three new RoRo ships, and these are still using diesel engines and scrubbers (delivered from a Chinese shipyard in 2020–2021, with a contract value of over 200 million. EUR). These new RoRo ships would have hybrid features in engines, since during port visits, vessels would use batteries and electricity for their main power. However, if LNG is as good solution as suggested in the press and research (especially in the long-term), then the abovementioned active companies in this regard clearly have an advantage during the following decade. It is also reported that Maersk is planning some LNG investments in the future due to 2020 demands (Zawadzki 2016). CMA CGM already placed initial orders for nine very large containerships at the end of 2017 (deliveries scheduled for 2020)— in the press release it is mentioned that these will easily fulfill the new sulphur cap demands of 2020 (CMA CGM 2017). However, it should be said that LNG is not a problem-free fuel. As it is new fuel, it is not so easy to find tanking places and, if they exist, how economically gas is supplied there. This could cause LNG price increases, especially if pipeline tanking is not available, and energy supply for the last mile is completed with trucks or ships (Schinas and Butler 2016). Pipelines and sea vessels are needed instead of trucks supplying gas. It is difficult to bring new fuel to the markets as it is not only about vehicles or vessels, but also fuel distribution system needs to be built in many places. In addition, the question of other harmful emissions of LNG, like methane, still needs to be resolved (Anderson et al. 2015; Baresic et al. 2018), even if emissions of CO_2, sulphur and nitrogen are in better control.

Ownership changes have also been present in the analyzed shipping companies in recent years. Company 'RoRo I' was gradually acquired by an Italian shipping group with an annual revenue of some billions of EUR. The acquisition process was rather a long one as it started in

2006, and ended ten years later (when they reached 100% control in the company). In June 2018, it was announced that one of the world's largest container shipping companies would be willing to acquire the analyzed company, 'Container.' In turn, company 'Bulk' of this analysis announced at the end of June 2018 that it would acquire a similarly sized Swedish shipping company (to double its size in revenue terms). These processes indicate that within the northern Baltic Sea shipping market and in the face of growing environmental pressure as well as slowly changing revenues, there is a tendency for shipping sector concentration to grow, and smaller actors to just disappear, while being merged to form larger entities.

5.3 Two Hinterland Transport Companies—Road and Rail

It could be assumed a priori that hinterland transport in Finland, for example, would have done better compared with sea transport. This is because short sea shipping was favoured to Estonia, and this would have required more hinterland transport and, in particular, road transport services. In this section, two Finnish-based hinterland transport companies are analysed, which are either specialized on road or rail. Both of these of course also have freight forwarding, terminal operations, warehousing, customs clearance services and so on available. The railway company is also active in road transport, but has emphasized over the years its railway involvement, and especially its focus on international eastern railway operations. It could be said that the road transportation company is more present in Finland, and within export and import flows to different European Union countries, while the railway company has a mixture of both east and west in its portfolio. Both of these companies are present in the Baltic States. These two companies are also both publicly traded.

As competition is intensive in hinterland transport, and particularly road transports in the European Union area (cross-border), it is not surprising to note that revenue development of the road transportation company has only been conservatively upwards (+43.9% in 8 years). Revenue did not experience that much change in 2015 (the year 2014 was only 9 months long in accounting terms, which shows growth, but no change in the in long-term), and still the best year for the company was 2013 (see Fig. 5.5). The railway-focused company in turn has gone

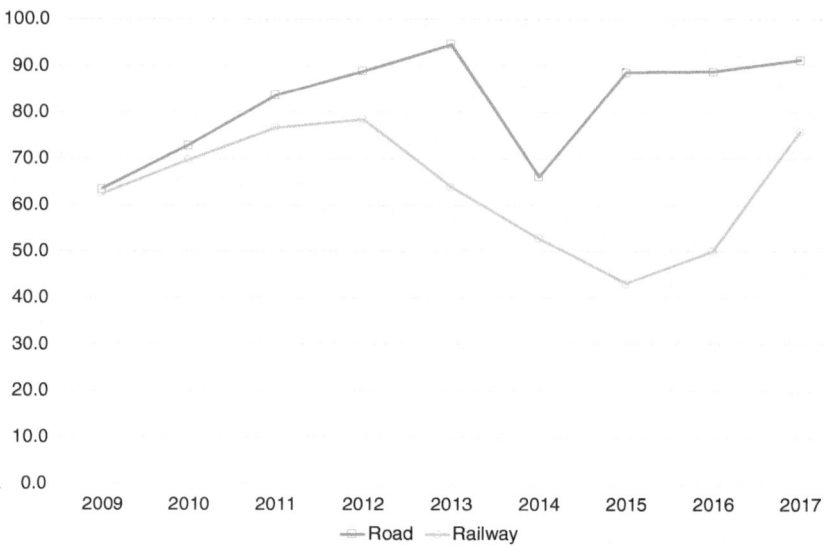

Fig. 5.5 Annual revenues of two Finnish hinterland transportation companies (million EUR) during the years 2009–2017 (year 2014 data: road transportation company accounting period length only 9 months, others 12 months) (*Source (data)* Annual reports, profit and loss statements)

through a rollercoaster ride in terms of revenue development, and the poor development of 2013–2015 was awarded with growth, and actually with quite substantial growth, in 2017. However, if examination is taken from year 2009 to 2017, it could be concluded that the growth of the railway company (+21.3%) has been half that of the road transportation company. Of course, the entire growth was produced in the last observation year, since with the 2016 data, longer-term development would be negative. Sulphur regulation did not have any positive effects on this company, and actually in 2015 its revenues continued to decline. Growth in 2017 was argued in the annual report to originate from Finnish export industries, and better import volumes (also, the logistics markets of Russia and the Baltic States were doing rather well).

Since competition at hinterlands is very intense, profits are typically low. This is the case in the two analyzed hinterland transportation companies (Fig. 5.6). Where road transportation company has been able to post some profits annually (apart from 2011), the profitability of the

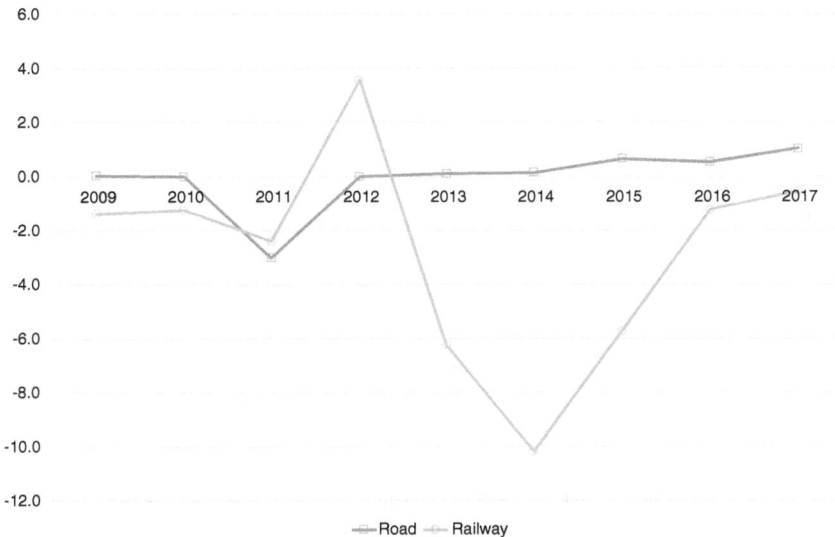

Fig. 5.6 Profit and loss of two Finnish hinterland transportation companies (million EUR) during the years 2009–2017 (year 2014 data: road transportation company accounting period length only 9 months, others 12 months) (*Source (data)* Annual reports, profit and loss statements)

railway company has been under great pressure. In fact, in the years 2013–2015 this company produced significant deficits, and it has been able to improve its performance over the last two years so that deficits are close to zero. Of course, Russian trade challenged (currency devaluation and even GDP decline in 2015) its transit and railway business, and export industries in Finland greatly suffered from the lack of growth in times of big deficits.

The amount of assets follow the changes of profit and loss—actually, the road transportation company has remained with total assets of around 20 million EUR, while the railway company has experienced significantly declining development to around 50 million EUR (Fig. 5.7). As losses increased substantially during the years 2013–2015, the railway company was forced to sell some of its assets (such as wagons used in Russian transports) and some of its business areas. They also gathered funds from investors to have the necessary working capital in operation. So, neither of these companies have expanded their asset base

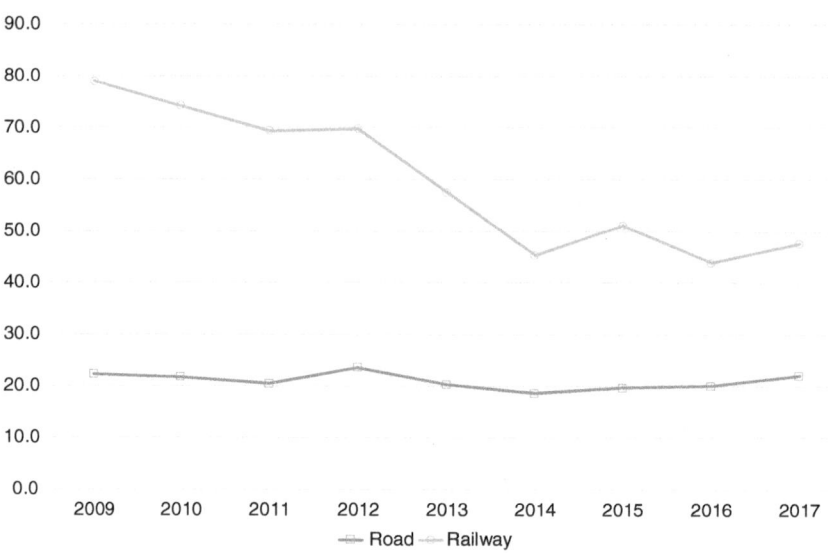

Fig. 5.7 Total assets of two Finnish hinterland transportation companies (million EUR) during the years 2009–2017 (*Source (data)* Annual reports, profit and loss statements)

in the observation period. The trucking company has reported that it has renewed its fleet during the years similar to the railway-focused company, but to a lesser extent.

Annual reports of both of the companies were also examined to assess whether they would have reported changes or investments due to the implementation of changes pertaining to sulphur regulation. It is interesting to note that of these two reported companies, neither one had any changes or challenges. The implementation time was not seen as challenging at all.

5.4   PRIVATE SECTOR EXPECTATIONS AND IMPRESSIONS BEFORE AND AFTER IMPLEMENTATION OF SULPHUR REGULATION

To give comprehensive picture of the sulphur regulation change, it is beneficial to go through qualitative interview studies before and after 2015, the year of implementing these changes. All of the interview studies reviewed in this section were carried out in Estonia and Finland.

The first qualitative study was completed in the autumn of 2011, and Finnish–Estonian interviews were a part of larger set of interviews concerning new railway alignment investment (there were also interviews completed in Latvia, Lithuania, Poland and Germany). However, in this study, the Finnish–Estonian interviews were reported as their own separate chapter (Henttu et al. 2011), and together with one Latvian private sector respondent, they were the only ones that commented on and that gave thorough opinions about the forthcoming change. The total amount of private sector respondents in the Finnish–Estonian sample was 15. They represented varying size of companies (in logistics flow volume terms).

Responses in 2011 study varied as to whether they were given from Finland or Estonia. In the former, the forthcoming sulphur regulation effects were seen rather negative, and in the bleakest scenarios, they were the major threat to foreign trade and macro-economy. In company-specific answers, the Finnish private sector actors were interested in new routes, which were seen as one of the few remedies for this situation (together with technical investment). At the more strategic level, longer sea journeys were questioned, and using the Baltic States to reach Europe was given full consideration (especially, if a viable and functional new railway connection was a possible investment in the future). Companies clearly identified that logistics costs would not decrease in the forthcoming years, however, price pressure was considered to be high at the time of interviews (within the aftermath of 2008/2009 economic crisis). In the case of the Estonian responses, the effects of sulphur regulation were considered to be much lower. In general, respondents were planning to tackle the forthcoming change by increasing the amount of road transportation to reach Central Europe—this would be the situation if sea transportation prices increased too much.

Just on the eve of the implementation of sulphur regulation, Alkhatib (2015) completed his M.Sc. thesis interviews to private sector companies concerning the changing environmental legislation. The main purpose of this thesis was to develop alternative and innovative shipping solutions between Helsinki and Tallinn. Alkhatib (2015) completed four industrial company interviews (one in Finland and three in Estonia) and these were just one small part of the overall study. In fact, the study included another 10 additional interviews, but they concerned this innovative new shipping concept in detail, discussing technology and operations related to it. The summary of these four interviews was such that industrial companies were starting to favour small- and medium-sized logistics service companies due to the price pressure (instead of global third-party logistics companies). However, companies were still operating and planning to operate with the earlier low inventory, just in time and high order frequency principles. The cost pressure was high, and final markets buying end products were not interested in paying a higher price for the overall delivered package. So, in a nutshell, companies were trying to keep the earlier methods and systems alive, but were keen to cut costs from everywhere.

After implementation of sulphur regulation, Lappalainen (2016) completed his M.Sc. thesis for the sea port of Helsinki, and built alternative future scenarios concerning possible handling volumes of different cargo groups. Part of the study included interviews with four key persons in the port and foreign trade-related logistics flows of Finland. The study's main emphasis was not that of sulphur regulation, but interviews contained political, economic, social, technological and ecological issues. Interviewees identified that ecological issues were becoming increasingly more important, especially in maritime sector. The negative business effects of sulphur regulation were identified as being low, and what would have been expected prior to the 2015 changes. LNG was seen, or identified, as a solution for many current and forthcoming challenges (such as nitrogen emissions). Interviews revealed that environmental regulation would increase the freight rates.

5.5   LNG: Sea Port Terminal Point of View

Some years before the implementation of sulphur regulation in 2015, it was greatly debated in European politics as to whether LNG would bring the necessary competition and security to energy markets, since LNG compression and use of LNG tanker ships would enable imports from the Middle East and North America. Some European countries also hold their own gas reserves, which are available for use with the new fracking technology. Using LNG will of course require its distribution and handling network because LNG needs to be changed as gas, or loaded to other vehicles as LNG. It is natural that such handling and import–export points of LNG are at sea ports. This enables ports to receive LNG imported by ships, and also gives fluid access for hinterland transportation modes to be used in distribution service. On the top of this, sea ports of course serve an important customer group—ships—by selling

LNG to them. Some sea ports might have also gas pipelines, so sea ports as a distribution point in gas handling is even more important.

The challenge with LNG distribution network is that it is not cheap to be build. One small-to-medium-sized LNG terminal will cost hundreds of millions to half a billion EUR. This is a lot of money, and at the Baltic Sea region, such construction projects have been supported by the EU and/or local country level subsidies. EU-based financial support has been most handsome for former Eastern European countries. Currently there are LNG terminals (at sea ports) in Poland, Lithuania and Finland. There are also plans to build such terminals in Estonia, Latvia, Germany, Sweden and Russia. So, wide interest for LNG is arising from different sources: the ability to meet environmental demands (of transport, but also heating and energy production), increases the competition in energy markets and lowers the price of gas.

From Fig. 5.8 the financial performance of the existing LNG terminal, which is a publicly traded company and is located in the southern part of Baltic States, can be accessed. The terminal also serves the oil

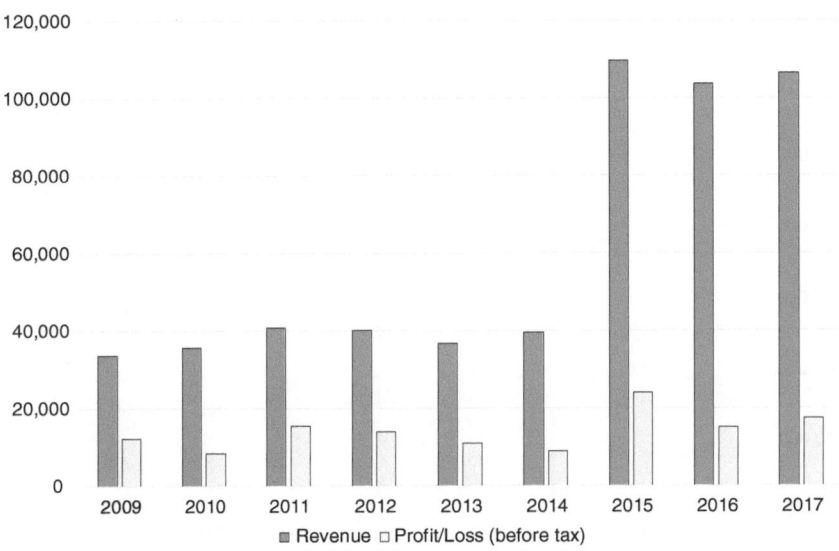

Fig. 5.8 Revenue as well as profit and loss development (thousand EUR) of sea port terminal operator located in the southern part of Baltic States during 2009–2017 (*Source (data)* Annual reports)

industry as a transit handling and storage point (for domestic companies, but also for foreign hinterland refineries). As Fig. 5.8 shows, oil handling was and still is a very predictable and profitable sea port terminal business. Before LNG implementation in the years 2009–2014 revenues, were approaching 40 million EUR, and profits before taxes varied at around 10 million EUR. With the introduction of LNG services in 2015 (and finalization of the main construction of LNG terminal), revenues increased considerably, and are currently above 106 million EUR. LNG business nowadays provides 64.1% of overall revenue. However, it is not as profitable as oil handling. This is of course understandable. Based on the company presentation, LNG provides around 30–40% from profits (Fig. 5.9).

As discussed previously, LNG investments have been supported by the EU and/or local governments, which is also the situation in this company. In fact, EU support was substantial for this investment as it was considered to increase energy security in the region (and as investment was made to the former Eastern Europe, which at the moment has an

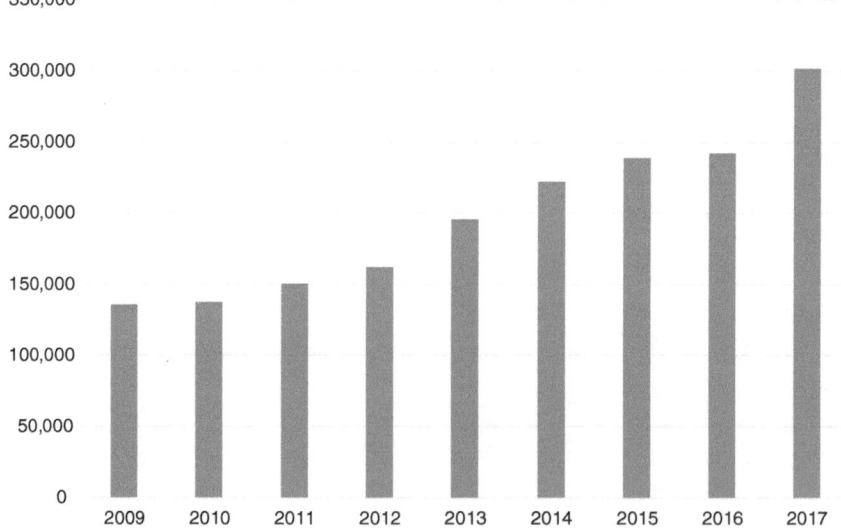

Fig. 5.9 Total assets of sea port terminal operator (thousand EUR) located in the southern part of the Baltic States during the period of 2009–2017 (*Source (data)* Annual reports)

emphasis on EU-supported investments, through policies of cohesion). Through examining the total assets development of this company, it could be detected that from 2011, the amount of assets have doubled (Fig. 5.9). Most of this invested money has benefitted LNG terminal construction and implementation. Despite being the first in the markets (in this geographical area), and receiving public financial support, from a company point of view, LNG investment has not been extremely profitable (although it has been good). For example, in the years 2009–2011 the company produced profits (before taxes) over all assets around 8.5%. In the years 2015–2017, this average produced profit to assets was 7.4%. Therefore, it could be concluded that without public support, this LNG distribution network point would be difficult, if not impossible, to build (with the current technical and economic realities).

This implemented LNG terminal, however, also has an impact on macro-economies, and local energy markets (Prontera 2017). It was argued that Lithuania received significant price reductions on gas prices from competing sources as the LNG terminal project was completed. These discounts were already so substantial that they had covered most of the investments of new LNG terminal. In addition, other Baltic States nowadays import gas through this new facility, which is providing them with the necessary multiple sourcing approach with cost and availability benefits.

It is difficult to say where and to what extent LNG terminal investments will be made, and how profitable these are going to be. It is quite dependent on the extent to which LNG will be used in transport, and particularly in shipping in the Baltic Sea region. The main problem in large-scale use is the lack of retrofits to ships, but also the non-existent secondhand market for LNG ships. Therefore, all investments need to be new builds, which require a lot of money. However, environmental emissions need to be cut—not only that of sulphur, but also nitrogen and CO_2. LNG is a good alternative to achieve these ambitious goals, but it is only one alternative among many (continuing to use old diesel or using methanol or even electricity). This alternative also has its weaknesses, and terminal infrastructure and the space necessary from ships are the most important ones. The most recent LNG report by IGU (2017) indicates that, globally, the biggest LNG-receiving terminal boom will become saturated in the time period leading up to year 2022. The current utilization of LNG-receiving terminals is between 30 and 40%.

5.6   Prices of Natural Gas in Different Locations

Since LNG is seen as a promising new alternative in the prevention of environmental emissions in maritime supply chains, as well as in an economic sense (lower cost fuel), it is therefore required to examine its prices further. Prices are ultimately at the heart of its future success, or failure, as a new fuel, and they are the key for further actions regarding new investments in ships, distribution infrastructure and possible enlargements of gas production. Gas has traditionally been used as a source of energy and heating within North America, former Soviet countries and some Central European countries. Asian countries are only at the start of wider gas use, and the highest growths in gas consumption in the following decades are forecasted for China, India and emerging Asia (BP 2018). Gas is very useful source of energy for heating (or cooling) as it does not pollute as much compared with coal or oil. This is especially beneficial for the air quality of mega-cities.

The price of gas varies among different continents, and as Fig. 5.10 illustrates, the price in North America has been contrarian to all price increases after the year 2009. The North American price has been down, and it is currently at the level of 1990s. This is a very rare development for any commodity, but it is based on two issues: (1) increased local production (due to fracking technology), and (2) it is expensive to export as distances to other markets are long (even in liquefied compressed form). Two other price developments—that of Asia (Indonesian LNG to Japan) and Europe (Russian gas to Germany) have shown much higher prices— and actually the strength in prices until late 2014 and early 2015. Price swings up and down are huge and resemble that of oil. In general, it could be concluded that the prices of gas have been low in the last years of observation period, and particularly low in North America.

If companies are seeking for remedies to avoid the upwards and downwards swings of the oil market with LNG, it is probably unlikely to happen. Average prices, standard deviation of these monthly prices and standard deviation percentage vs. average were calculated from these three types of gases, and two types of oil. Standard deviation as a proportional measure (%) is 63% from the period of January 1992–June 2017 for imported gas to Germany and Japan. It is 57% for North American gas. This same price fluctuation measure is 69% for Brent and 67% for the simple average of three different prices of oil. Therefore, in retrospect, it could be stated that only in North America has gas been the

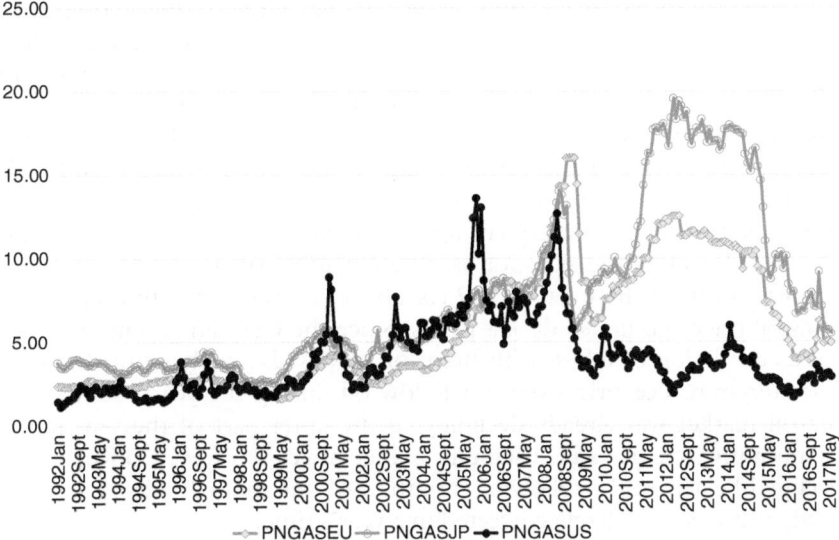

Fig. 5.10 Monthly prices of natural gas (USD per million metric Btu) in three different locations (denotations: PNGASEU: Russian natural gas border price in Germany; PNGASJP: Indonesian Liquefied Natural Gas in Japan; PNGASUS, Natural Gas spot price at the Henry Hub terminal in Louisiana) during the period of January 1992–June 2017 (*Source (data)* IMF 2018)

'safer' source of energy, but German- and Japanese-imported gas is nearly as high in its price fluctuation as oil.

Of course, it could be argued that examining gas prices in longitudinal form since early 1992 until June of 2017 is not fair, since its wider use has only taken place during the last decade. If average prices, standard deviation of these prices and standard deviation (%) is calculated with data starting from January 2008, the results remain mostly the same. Standard deviation of price is much lower in all five types, since prices in general were higher in this time period—German and Japanese gas prices as well as both oil prices have the same fluctuation, around 32–36%. Only North American prices have changed more as standard deviation (%) is 48%. This latter fluctuation has been to the benefit of the buyer (and eventually consumer) as prices have declined in the period very significantly (also concluded in Thiel and Masters 2014).

These five different energy sources also have a fairly high correlation in prices with each other, apart from North American gas. Table 5.1 illustrates this situation further, where North American price correlations are shaded with gray. So, the North American gas price has not followed, or has followed with in a varying manner, the prices of the other two gas-importing countries and the prices of oil. It is simply the case that North American gas has been cheap and low priced since the 2008 peak. It has shown very small and infrequent upward movements in price over the last decade. However, the situation is different in German and Japanese gas imports. Both of these have a 0.9 (or above) correlation with oil prices—so as oil prices go up, so do gas prices. Since the German gas import price has a correlation of 0.898 with Brent oil, it was also analysed further with a scattergram. Gas prices did not follow oil during the year 2008: When the oil market was already declining in the latter part of the year rather significantly, was gas still sustaining higher prices. However, prices of gas declined heavily in the following months. So, this anomaly was very short-lived, and was completely corrected in a year's time.

In the literature, there has been a lot of talk about the favourable price difference of LNG over diesel oil. This was also tested with available longitudinal data and with conversion of million metric BTU (gas) to mega joules, as well as making the same conversion from barrel of oil. In the longer-term (with data from the entire observation period, starting from 1992), North American gas has indeed been much cheaper than Brent oil—40.6% cheaper. Of course, this difference varies over time, and in some months, gas in North America could have cost double the price of

Table 5.1 Correlation between monthly gas prices and prices of two oil qualities at three different locations

	PNGASEU	*PNGASJP*	*PNGASUS*	*POILBRE*	*POILAPSP*
PNGASEU	1				
PNGASJP	0.872	1			
PNGASUS	0.429	0.238	1		
POILBRE	0.898	0.930	0.418	1	
POILAPSP	0.900	0.922	0.435	0.999	1

Source (data) IMF (2018)
Denotations: POILBRE: Brent quality USD price per barrel; POILAPSP: average USD price of Brent, West Texas Intermediate and the Dubai Fateh

oil, but in very favourable circumstances it has been 90% cheaper. Being lower cost than oil is also supported with price of gas imported to Germany from Russia. During the observation period, it was 27.4% cheaper compared with the same energy content of Brent oil. However, here fluctuation between is as steep as it is in North America—in some months gas in Germany has cost below half of the Brent oil price, but at other times, it has been more than double the cost of oil. In the case of Japanese gas imports from Indonesia, it is the case that price favourability of gas drops quite a lot. Gas is on average 2.8% lower priced, however, with a similar fluctuation to that reported earlier (e.g., some months it could be 40% cheaper, but in some 80% higher compared with the same energy content price of oil). What is the relevance of these diverse results for this book? For North America and Germany, LNG is brought to markets mostly with the most efficient transportation solution—that of pipelines. Japanese markets are served with LNG ships, which compress gas to six hundred times in the port of departure and then these are exported to Japan for storage and use. The future of European LNG price (and that of the Baltic Sea) will depend on a mixture of Japanese and German practices, and prices will follow these. If the history of these markets is any proxy for future development and the wider use of LNG in ships in the Baltic Sea, it could be assumed that prices will be lower as compared with Brent oil, but they are not going to be one fourth lower. It is quite possible that the long-term average discount will be around 15–20%. In the case of a higher amount of LNG imported from other continents by ships, it could be the case that the price advantage is only 10%. However, the situation in North America is different. As natural gas is low priced in this region, and has remained so (also after observation period of this analysis; EIA 2018), it would not be that surprising to see growing interest for LNG-powered ships for the fleet serving this area. The situation is the opposite in Asia, where natural gas is considered to be expensive, with nearly the same price as oil (in the highest demand season, it may even be much higher; see Jaganathan 2018). Quite a lot depends on LNG use, how supply from North America is arranged in cost and transportation capacity terms to Asia, and particularly to China (Raju et al. 2016; Livsey 2017).

5.7 Conclusions

The change of emission regulation in 2015 within the Baltic Sea region brought changes to the maritime supply chain actors, and many companies have experienced this change as only the beginning of the more demanding era. It is a question of time as to when nitrogen and CO_2 emissions will be asked to be reduced (or prevented entirely, in the case of nitrogen), and they are actively followed. Therefore, it is understandable that shipping companies reacted to the changes of 2015 mostly with a 'wait and see' approach by using low sulphur content diesel oil. Only one company completed a significant investment program, and invested in scrubbers within their ships. However, many companies have been active in the LNG front, and they either have these in use or they will receive new LNG ships from shipyards for service in the near future. LNG currently has a clear advantage over diesel oil in emission terms, and with many pollutant measures, and it is also still clearly a cheaper fuel than oil. However, there is a lot of uncertainty about the future of LNG. It requires investments in new ships (as the secondhand market does not exist) and new LNG terminals. Ships have higher costs than conventional diesel-powered ships, and LNG terminal(s) require the investment of hundreds of millions of euros. As LNG terminals that already have investment (with such investment being financially supported by EU) show, these huge terminal investments are not necessarily extremely profitable for terminal operators, but they can change energy dynamics and local energy markets in positive way. It may be the case that LNG transition needs to be seen from a wider perspective. However, as threat for the future, is the price uncertainty of gas, and particularly the possible higher price of imported LNG. The price benefit over oil is not necessarily one fourth, and it could be much smaller. Based on a simulation study of Kana and Harrison (2017), the price difference between natural gas and diesel oil is at the core of a wider adaptation of LNG shipping.

Hinterland transportation companies did not benefit from demanding sulphur regulation. Two companies were analysed in this chapter, and the road transportation company seemed to have gained some conservative revenue growth during the years, but profits remained small throughout the examination period. Another company more concentrated on east (Russia) and railways experienced a very difficult time period from 2013 to 2015, in which it reported rapidly declining revenue development and significant losses. It could be the case that sulphur regulation change was not big enough (and oil prices were very weak in 2015–2016) in order for hinterland transport to gain significant advantages over sea transport. In addition, competition is very intensive in hinterland transportation (e.g., Kummer et al. 2014), and the competitiveness of Finnish companies might not be as high as compared with Polish- or Czech-based competitors (e.g., in road transports).

References

Alkhatib, A. (2015). *Developing short sea shipping transportation chains at Helsinki-Tallinn route*. Unpublished M.Sc. thesis, Lappeenranta University of Technology (LUT), Lappeenranta, Finland. Available at http://urn.fi/URN:NBN:fi-fe201502201663. Retrieved 17 May 2018.

Anderson, M., Salo, K., & Fridell, E. (2015). Particle—And gaseous emissions from an LNG powered ship. *Environmental Science and Technology, 49*(20), 12568–12575.

Baresic, D., Smith, T., Raucci, C., Rehmatulla, N., Narula, K., & Rojon, I. (2018). *LNG as a marine fuel in the EU. Market, bunkering infrastructure investments and risks in the context of GHG reductions*. London: UMAS.

BP. (2018). *BP energy outlook 2018*. Available at https://www.bp.com/en/global/corporate/energy-economics/energy-outlook.html. Retrieved 18 March 2018.

Burel, F., Taccani, R., & Zuliani, N. (2013). Improving sustainability of maritime transport through utilization of liquefied natural gas (LNG) for propulsion. *Energy, 57,* 412–420.

Churchill, J. (2017, September 13). Switch, scrub, or just cheat? *A.P. Moller—Maersk*. Available at https://www.maersk.com/stories/switch-scrub-or-just-cheat. Retrieved 7 April 2018.

CMA CGM. (2017, November 7). *World innovation: CMA CGM is the first shipping company to choose liquefied natural gas for its biggest ships*. Press release of CMA CGM. Available at https://www.cma-cgm.com/news/1811/world-innovation-cma-cgm-is-the-first-shipping-company-to-choose-liquefied-natural-gas-for-its-biggest-ships. Retrieved 27 June 2018.

EIA. (2018). *Henry Hub natural gas spot price*. US Energy Information Administration. Available at https://www.eia.gov/dnav/ng/hist/rngwhhdd.htm. Retrieved 28 May 2018.

Gritsenko, D. (2018). Explaining choices in energy infrastructure development as a network of adjacent action situations: The case of LNG in the Baltic Sea region. *Energy Policy, 112,* 74–83.

Henttu, V., Terävä, T., & Hilmola, O.-P. (2011). Results of Finnish and Estonian private sector interviews (LUT Kouvola), chapter 6. In *Private transport market stakeholders in the area of Rail Baltica*. City of Warsaw, Poland: EU-Consult.

IGU. (2017). *2017 World LNG report*. Barcelona, Spain: International Gas Union. Available at https://www.igu.org/sites/default/files/103419-World_IGU_Report_no%20crops.pdf. Retrieved 19 March 2018.

IMF. (2018). *IMF primary commodity prices*. Available at http://www.imf.org/external/np/res/commod/index.aspx. Retrieved 19 March 2018.

Jaganathan, J. (2018, January 19). Asia spot prices climb to three-year high on winter demand. *Reuters*. Available at https://www.reuters.com/article/us-global-lng/asia-spot-prices-climb-to-three-year-high-on-winter-demand-idUSKBN1F825V. Retrieved 28 May 2018.

Kana, A. A., & Harrison, B. M. (2017). A Monte Carlo approach to the ship-centric Markov decision process for analyzing decisions over converting a containership to LNG power. *Ocean Engineering, 130,* 40–48.

Kummer, S., Dieplinger, M., & Fürst, E. (2014). Flagging out in road freight transport: A strategy to reduce corporate costs in a competitive environment—Results from a longitudinal study in Austria. *Journal of Transport Geography, 36,* 141–150.

Lappalainen, J. (2016). *Helsinki hub—New traffic flows and business networks*. Unpublished M.Sc. thesis, Lappeenranta University of Technology (LUT), Lappeenranta, Finland. Available at http://urn.fi/URN:NBN:fi-fe201603038503. Retrieved 17 May 2018.

Lindstad, H. E., Rehn, C. F., & Eskeland, G. S. (2017). Sulphur abatement globally in maritime shipping. *Transportation Research Part D, 57,* 303–313.

Livsey, A. (2017, December 6). LNG tankers: Cooking with gas. *Financial Times*. Available at https://www.ft.com/content/ffbd0b0e-da88-11e7-a039-c64b1c09b482. Retrieved 28 May 2018.

Line, Stena. (2016). *Annual review—Stena AB*. Sweden: Gothenburg.

Ma, H., Koen, S., Xavier, R. P., & Nigel, T. (2012). Well-to-wake energy and greenhouse gas analysis of Sox abatement options for the marine industry. *Transportation Research Part D, 17*(7), 301–308.

Olaniyi, E. O. (2017). Towards EU 2020: An outlook of SECA regulations implementation in the BSR. *Baltic Journal of European Studies, 7*(2), 182–207.

Prontera, A. (2017). *The new politics of energy security in the European Union and beyond: States, markets, institutions*. London and New York: Routledge and Taylor and Francis Group.

Raju, T. B., Sengar, V. S., Jayaraj, R., & Kulshrestha, N. (2016). Study of volatility of new ship building prices in LNG shipping. *International Journal of e-Navigation and Maritime Economy, 5,* 61–73.

Schinas, O., & Butler, M. (2016). Feasibility and commercial considerations of LNG-fueled ships. *Ocean Engineering, 122,* 84–96.

Terazono, E., & Hume, N. (2017, May 31). New shipping fuel regulation set to hit commodities. *Financial Times*. Available at https://www.ft.com/content/d0ae63c4-452f-11e7-8519-9f94ee97d996. Retrieved 7 April 2018.

Thiel, P., & Masters, B. (2014). *Zero to one—Notes on startups, or how to build the future* (1st ed.). London, UK: Virgin Books.

Yang, Z. I., Zhang, D., Caglayan, O., Jenkinson, I. D., Bonsall, S., Wang, J., et al. (2012). Selection of technologies for reducing shipping Nox and Sox emissions. *Transportation Research Part D, 17*(7), 478–486.

Zawadzki, S. (2016, February 22). Qatar, Maersk and Shell join forces to develop LNG as marine fuel. *Reuters.* Available at https://de.reuters.com/article/shipping-lng/qatar-maersk-and-shell-join-forces-to-develop-lng-as-marine-fuel-idUKL8N1611S1. Retrieved 7 April 2018.

Zis, T., North, R. J., Angeloudis, P., Ochieng, W. Y., & Bell, M. G. H. (2015). The environmental balance of shipping emissions reduction strategies. *Transportation Research Record: Journal of the Transportation Research Board, 2479,* 25–33.

CHAPTER 6

Longitudinal Survey Findings from Northern Europe

Abstract Numerous logistics flow direction and warehousing surveys have been completed in Finland and Sweden during the period of 2010–2015. Surveys have been conducted with the largest companies in these countries (excluding banking, finance and the service sector). In 2015, Estonia's largest companies were also included in these surveys. Based on these surveys, it could be concluded that the most typical transportation unit in companies is semi-trailer/trailer, followed by container, and other units. Companies did not see much change in the mode of transportation used: the surveys indicate that road transport will dominate in the future, followed by sea transport. Most of the companies indicated that tightening environmental legislation will increase their transportation costs and the survey of 2015 clearly showed that Finnish companies were most hurt by the implementation of strict sulphur regulation.

Keywords Transportation units · Transportation modes
Environmental legislation change · Northern Europe

6.1 Introduction

Supply chains and their transportation modes as well as transportation units differ greatly between different geographic regions. Typically, in Asian countries it is popular to use containers—in sea transport, but also at hinterlands (since weight restrictions are so tight).

© The Author(s) 2019
O.-P. Hilmola, *The Sulphur Cap in Maritime Supply Chains*,
https://doi.org/10.1007/978-3-319-98545-9_6

However, hinterland transport in unitized cargo in Asia is still completed almost solely with trucks (Shibasaki and Watanabe 2010). This is actually a smart strategy concerning the future in a situation if or—in fact—when environmental emissions begin to be followed, and reductions are sought (containers could be loaded on trains then, instead of trucks). Containers are also popular in North America—Asian trade and the North American side have even developed their own set of special containers, which are much longer than elsewhere, and these could be transported very efficiently at maritime–railway supply chains. In North America, containers are also stacked together at railways (which is not done in Europe as railway electricity networks prevent this sort of technique from be applied), and trains are more than 2 kilometers long (again, a standard that is difficult, if not impossible, to match in Europe).

Europe is in many respects different from the other areas. The continent is very keen to be at the forefront of emission reduction, sustainability and green values. However, its logistics sector, or functioning daily supply chains, hardly follows these principles particularly deeply. The European Union is still a rather young, larger economic area, and many small nations have been keen to keep their railway sector and unions happy by opening sector for free competition in slow motion. European countries also typically have hinterland access to most of the places, where the main transportation needs to exist. Therefore, trucks are very often used—trucks accompanied with semi-trailers. This combination is most interoperable around European countries and could be integrated swiftly to maritime transport with RoPo/RoPax ships. Of course, emissions are a great problem here, but these have not been a major issue in public debates so far. It is somewhat daunting that . nearly weekly (during 2017 and 2018) there was a debate about private cars, diesel, emissions, and prohibiting diesel car use (or both diesel and petrol) in some countries (or city; e.g. Boston 2018; Pullella 2018; Bennett and Vijaygopal 2018). Some car manufacturers have even announced that they will stop producing diesel cars, and limit the production of higher emitting cars. This discussion is very rarely taken to the supply chain level, and to concern industrial as well as commercial use of diesel in chains, and particularly that of trucks with semi-trailers. Everyone knows that making this combination more liable for different kinds of emissions would be a hurdle for centralized European supply chains (Bershidsky 2018 touched this issue). It would be very difficult to implement low inventory and just-in-time supply chains otherwise.

Trucks dominate transportation mode statistics if transportation mode shares are measured with tons, instead of with the two dimensional ton-km. Latter mentioned shows some prospect for a sustainable future. For example, in Germany, a country with a long railway tradition and the largest single railway freight in the EU country, road transportation has a share of 65.2%, when measured with ton-km, but this share increases to 83.8% if tons are the measure (based on data of European Union 2018 for year 2015). The same applies to Sweden (European Union 2018), which has an abnormally high railway freight market share from the overall market (year 2015 situation): if the freight market is measured with ton-km, the railway modal share is 33.3%, and with tons it is 13.3%. Correspondingly, in Sweden, hinterland dominance of roads increases as railway freight loses its modal share (from 66.7 to 86.7%). Finland is very similar to Sweden—the road has a modal share of 74% with ton-km and 88.92% with tons. Only Estonia, as a small country with a proportionately high amount of freight compared to its size (some energy raw materials and transit), sustains high railway modal share whatever the measure is (in ton-km it is 33.2% and in tons it is 49.9%).

This chapter will review the results of different surveys, which were completed for Finnish and Swedish companies. These respondents were selected in each year from the largest company lists, and they were contacted to respond via email and Internet questionnaire. The survey offered the opportunity for response inlocal languages (in Finland: Finnish and Swedish, in Sweden: Swedish), and in English. Surveys reported in this chapter are from years 2010, 2011, 2012 and 2015. The survey was also completed in years 2006 and 2009, but due to their older appearance, they are not included in here. In the 2015 survey, the largest Estonian companies were also included as part of research, and the survey was offered in both Estonian and English. In this section, transportation units used are analysed, followed by transportation modes and finally issues related to sulphur regulation are analysed. All of these responses are from companies, and in earlier studies respondent background was considered to be good (due to the quality of responses since respondents have a long experience of company operations and logistics). The amount of responses is not particularly high, but these responses will shed some light on and some direction for the North European situation. In addition, the longitudinal perspective provides further reliability since the amount of responses is from different years (summed together), and together the amount is fairly good one.

6.2   TRANSPORTATION UNITS USED

The leading and largest companies of Estonia (year 2015), Finland and Sweden (years 2010, 2011, 2012 and 2015), were asked what their most important transportation unit is in the supply chain (financial, insurance and service sector companies were excluded). Response rates of this survey were from 3.5 to 5% depending on the survey year, so contact email to encourage response to the survey varied from more than 500 to somewhere above 800. The survey was longer one, and it contained all sorts of questions concerning transportation flows, amounts and warehousing issues. Transportation units used was just one part of the response, and in this section only analyzes those responses that were valid, and contained the complete information that was requested. Responses could be tracked to companies, and partially to the respondent persons (if they did provide this data). It could be said that respondents represented large companies, and they had significant experience of logistics and supply chain management.

Responses in the years 2010–2015 clearly indicated that the supply chain strategy of Finnish and Swedish companies is tied to the large-scale use of semi-trailers/trailers (Fig. 6.1). As there are only few responses from Estonia, and these are limited to the single year of 2015, it could be said that the data suggest that the same situation is taking place over there. However, it seems reasonable to conclude that around 60% of companies base their supply chain on trucks and semi-trailers (or some kind of trailer combination). Only in 2012 does this share drop to 53.6%, as a class of 'something else' takes an abnormal share in that year of responses.

It is quite interesting to note that companies do not see containers as their most important transportation unit as only around 20% identify their importance (in the first survey year it is 24%). This container use is an equal mixture of 20 and 40/45 foot containers. The type of container used is of course dependent on cargo type, customers and supplies. There is a preference for shorter containers for heavier cargo (since the payload is high compared with space), whereas consumer goods in turn require space rather than weight. Do note that in the 2012 responses, containers were used as a transportation unit only for 10.7% of respondents.

There are also somewhere above 20% of companies (in 2012 it was abnormally high at 35.7%) that argue that something else is the most important transportation unit. These must be companies with either raw

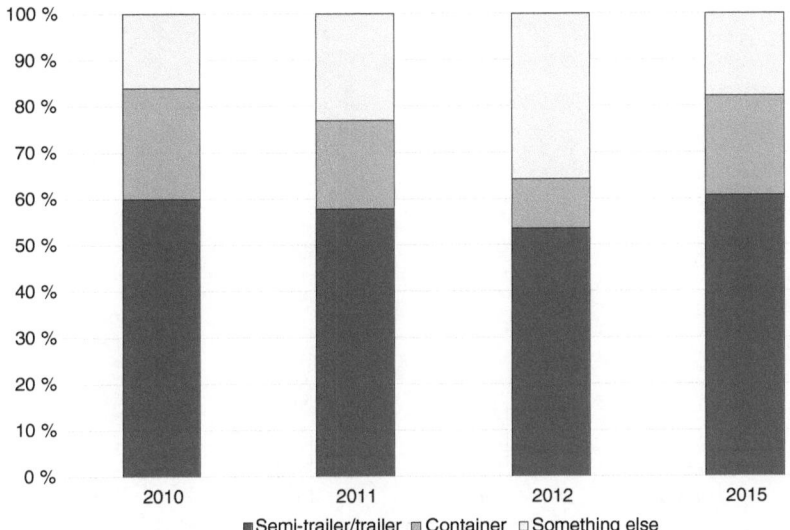

Fig. 6.1 Transportation units used in larger companies in Finland (all years 2010, 2011, 2012 and 2015), Estonia (year 2015) and Sweden (all years) in years 2010 ($n = 25$), 2011 ($n = 26$), 2012 ($n = 28$), and 2015 ($n = 28$)

material or very low value added items to be transported, or alternatively, they may only have parcel deliveries, since price to weight is so high.

6.3   Transportation Modes Used: Future Directions

Due to the high preference for the use of semi-trailers/trailers and truck use, it is not surprising that the road is the most used transportation mode. This was still the case in the 2015 survey. Of course, this is company dependent, where some companies mostly utilise the railways or air transport, whereas others use road and sea transport.

It is interesting to note that companies in 2015 were still thinking that in the years 2017 and 2022, the road would dominate. In Fig. 6.2, the realized modal share in the year 2014 as estimated by respondent companies was 64.1%; this was expected to decline to 61.3% in the year 2022. Together with this, it is also notable that railways only have a share of a few percent. The railway share is growing somewhat, but it remains the lowest used mode of transport. This could be explained by the fact that

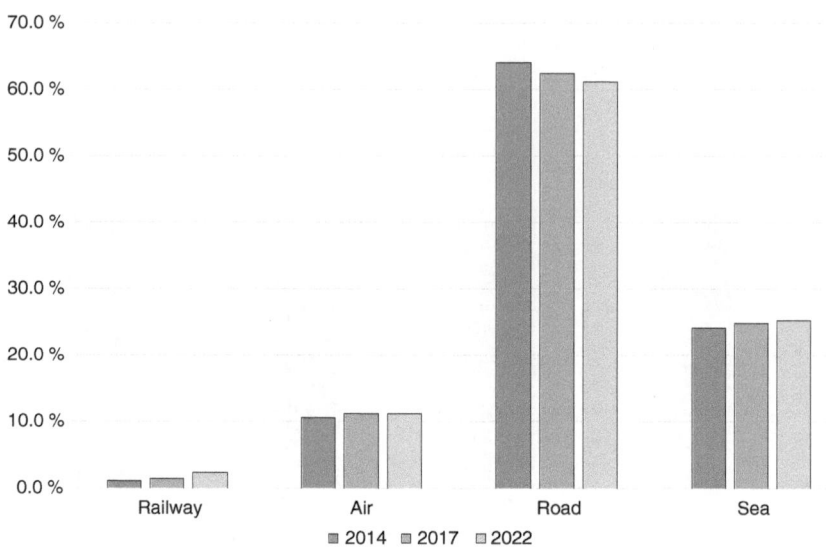

Fig. 6.2 Transportation modes used (percent share from total) in the 2015 survey during years 2014 (realized), 2017 (forecast) and 2022 (forecast)

railways are mostly used in respondent countries for raw materials and the transport of low-cost items as well as on transit cargo (in this last item could be carrying more valuable cargo, and containers). If respondents do not represent large enough companies, then the rate could be as low as shown in Fig. 6.2.

For both air transport and maritime transport modes, respondents forecasted some small growth during the forthcoming years. It is important to note that the 2015 survey was completed after the implementation of sulphur regulation, and companies already had early experiences from its effects (of course, oil prices were coming down at the time of the survey). The question of air or maritime transport is vital for some companies—it clearly dominates in some responses, and road transport comes only as the second consideration. The preference for air transport is understandable in sectors where the price of product per weight/volume is high, where products may deteriorate fast (fashion, high tech, food, pharmaceuticals etc.), and where markets are long-distance (e.g., global, and manufacturing in few locations). Maritime transport is typically used on a large-scale for products that are low value-added, and heavy.

6.4   Sulphur Regulation Effects in Two Different Surveys

Regulation change concerning maritime transport and sulphur regulation was dealt with in two different survey rounds. In the 2012 survey, Finnish and Swedish companies were already asked whether implementation of sulphur regulation in 2015 would cause an increase in transportation costs. Question included also potential increase of CO2 payments together with sulphur regulation (CO_2 payments are still under development for transportation in EU area). In total, 27 companies responded to this question, and 78% of these indicated that costs would increase. So, this result was rather clear to interpret, and supports earlier arguments and findings.

Later on, in the 2015 survey, respondents were asked what kind of effects sulphur regulation had on transportation costs. The responses were recorded in April 2015, so it was the correct moment for respondents to speak from some months of experience of the cost effects. As oil price was dropping in 2015, it was evident that the cost effects were not as harmful as they could have been. Interestingly, only three responding companies reported "no change" at all. The rest of the companies reported a cost increase (in total, 25 companies; Hilmola et al. 2017). Five companies reported cost increases of 4–6% or even above 8%. A priori to this and based on all the research, it could have been assumed that Finnish-based companies had most significant cost effects (of course these companies are multinational corporations, and operations could be all over, but having response from particular country is still good indication of operations having role in same country too). From these five companies, four were Finnish and one was Swedish. Most of the companies that responded that there was quite a substantial increase were typically producing something heavy, which required a lot of maritime transport. Alternatively, they were using maritime transport on an extensive scale (e.g. imports from Asia).

Since Estonia and Sweden do have better hinterland positions, and could use alternative hinterland routes to reach Central Europe, for example (instead of a longer maritime route), this again supports the theory and previous research, that the "no change" respondents arose from Estonia (two responses) or Sweden (one response). Based on Fig. 6.3, even the category of "2–4% growth" was over-presented by Finnish companies. So, based on this survey, sulphur regulation had the greatest effect on Finnish-based operations. Distant, isolated and maritime-dependent is a poor combination in light of the forthcoming global change in 2020. It is important to note that the maritime sector was in poor shape globally in 2015, and freight prices in most places in the world declined. United Nations (2016: 52) described container market prices in 2015 with phrases such as "*declined steadily*" and "*reaching record low prices*"—rates from Shanghai (China) to Europe declined as much as 45.82%, and from Shanghai to the West Coast of the USA by 23.55%. The decline in container freight rates continued in 2016 (United Nations 2017), but this showed some weak recovery too (like in the Shanghai–Europe route). The United Nations (2017) reported container freight rates to have developed in a negative direction in most of the globally followed shipping routes. Global freight index called the 'Baltic Dry Index' showed weakening from early 2014 to early 2015 by 31% (Bloomberg 2018); the decline continued in 2016 by 50% (from 2015 levels).

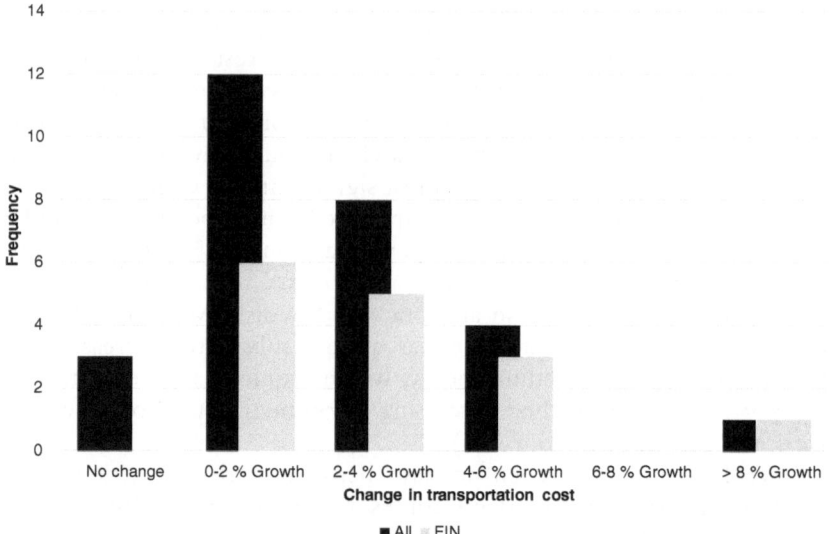

Fig. 6.3 Change in transportation costs due to implementation of sulphur regulation in respondent companies during the 2015 survey ($n = 28$), where the black column includes all responses (Finland, Estonia and Sweden), and lighter grey includes the Finnish (FIN) responses only

In an earlier longitudinal case study research from heavy industry representative of Finland, one paper mill showed that sulphur regulation was a big threat to sustaining gross margins and profitability (Hämäläinen et al. 2016). It was the case that distribution costs were already experiencing a difficult time period: demand for paper started to mature (and in some grades even decline) after the early 2000s, and simultaneously distribution costs were increasing considerably nearly every year (mostly increasing due to higher priced oil until 2009). It was rather challenging for low profit margin heavy industries that sulphur regulation did not provide any upsides (or if it did, they were very limited) in the case of a possible decline in oil prices. However, the downside was considerable, and had the potential to increase with the time since oil prices increase over the years and distribution needs continue (Hämäläinen et al. 2016). Based on FAO (2018) statistics, paper production (printing and writing paper) declined in Finland from year 2004 (an all-time high) to the year 2016 by 45% or by 4.26 million tons. In 2015, it declined by 2.8% from the previous year, and in 2016 annual decline accelerated to 7.8%. Compared with Sweden, same grade paper production reached its highest level in 2012, but it has declined annually ever since. Decline from year 2012 to 2016 has been 24.5% (or 0.84 million tons). Yearly declines accelerated in 2015 and 2016 by 6.6 and 9.5%, respectively.

6.5   Conclusions

European countries, and also Northern Europe in general, are very dependent on the use of trucks in supply chains. It is very popular to use a combination of truck and semi-trailer in cross-border deliveries as it will fulfill standards in each country for weight and length. Interoperability is then very high. Semi-trailers could always be just a platform, where they carry containers (like forty-foot equivalent unit). By analysing the survey results, mostly completed in Finland and Sweden, in the years 2010, 2011, 2012 and 2015 (only in the last year was Estonia also included), it could be concluded that road transportation still holds dominance in supply chains, and based on responses regarding the future, road transportation will be slow to lose its share to others. Therefore, it is easy to understand why truck and semi-trailer-based maritime supply chains with very short sea journeys have been popular. This is especially the case after the implementation of sulphur regulation in 2015. It was a very logical choice to proceed with, since one component in the supply chain became higher cost (or possessed the option to be and did not show any decrease when oil markets declined significantly in 2015). Of course, these supply chains are not sustainable or CO_2-competitive by any means. It is just the case that centralized, very low inventory just-in-time supply chains require road transport, and very fast maritime transport at a high frequency. It is also one of the few ways to keep delivery reliability in tight schedules—containers are always handled in container ship-based maritime supply chains numerous times by different parties, and the system is prone to errors, strikes, and subject to the availability of resources and technical problems.

In surveys of 2012 and 2015, companies were confident that higher demands on environmental regulation would bring about transportation cost increases. Companies also reported higher costs during the sulphur implementation year, and what makes this finding interesting is the fact that in the years 2015 and 2016 freight rates were globally very depressed. Finnish companies were mostly complaining about being hurt, which is logical since proper hinterland connection to Central Europe is lacking, and the maritime component is present in nearly every higher volume supply chain.

REFERENCES

Bennett, R., & Vijaygopal, R. (2018). An assessment of UK drivers' attitudes regarding the forthcoming ban on the sale of petrol and diesel vehicles. *Transportation Research Part D, 62,* 330–344.

Bershidsky, L. (2018, February 28). The end of diesel is here: Germany isn't ready. *Bloomberg View.* Available at https://www.bloomberg.com/view/articles/2018-02-28/germany-s-diesel-decision-bans-are-allowed-but-are-they-wise. Retrieved 28 March 2018.

Bloomberg. (2018). *Baltic Dry Index (BDIY: IND).* Available at https://www.bloomberg.com/quote/BDIY:IND. Retrieved 27 March 2018.

Boston, W. (2018, February 28). Diesel takes hit in German court—In blow to car makers, ruling clears way for cities to ban certain vehicles to cut pollution. *Wall Street Journal.*

European Union. (2018). Eurostat database. *Transport.* Available at http://ec.europa.eu/eurostat/web/transport/data/database. Retrieved 26 February 2018.

FAO. (2018). *Forestry production and trade statistics.* United Nations, Food and Agriculture Organization. Available at http://www.fao.org/faostat/en/#-data. Retrieved 25 April 2018.

Hilmola, O.-P., Kiisler, A., & Hilletofth, P. (2017). Cabotage and sulphur regulation change: Cost effects to Northern Europe. *International Journal of Business and Systems Research, 11*(4), 417–428.

Hämäläinen, E., Hilmola, O.-P., & Tolli, A. (2016). North European export industry and the shadows of sulphur directive. *Journal of Transport and Telecommunication, 17*(1), 9–17.

Pullella, P. (2018, February 28). Monument-filled, traffic-clogged Rome to ban diesel cars by 2024. *Reuters, Business News.* Available at https://www.reuters.com/article/us-germany-emissions-rome/monument-filled-traffic-clogged-rome-to-ban-diesel-cars-by-2024-idUSKCN1GC1DD. Retrieved 28 March 2018.

Shibasaki, R., & Watanabe, T. (2010). A comparison of semi-trailer transport of international maritime container cargo in Japan and South Korea, and its implications. *Procedia—Social and Behavioral Sciences, 2,* 6118–6129.

United Nations. (2016). *Review of maritime transport.* Geneva: Unctad, United Nations.

United Nations. (2017). *Review of maritime transport.* Geneva: Unctad, United Nations.

CHAPTER 7

Simulation of Different Supply Chain Strategies

Abstract Different future scenarios can be evaluated using computer simulation. This chapter illustrates that changes in the northern Baltic Sea region due to sulphur regulation were driven by speed and other cost items rather than purely freight cost. If the cargo has some value, price erosion and cost of inventory holding represent significant financial burdens in supply chains. These short lead time smart solutions do not experience significant problems if CO_2 emissions are put under some payment scheme (since distances are not excessive, even if trucks are used). However, for the global 2020 change and within continental supply chains, speed is not alone the answer. Air freight supply chains emit so much CO_2 that payment schemes would have a significant impact upon these chains. An alternative would be to use more hinterland transport (railway) between Asia and Europe and this seems to be a lucrative suggestion—it emits low levels and is also relatively fast since it also reduces total costs. Container shipping supply chains are in a difficult position since their lead time is already high and causes harm to cargo owners. This chapter also incorporates a simulation model to assess sulphur regulation cost effects, in which different factors can be taken into account to analyse what the impacts might be on freight prices.

Keywords Simulation · Supply chains · Price erosion · Inventory holding costs · Diesel costs

© The Author(s) 2019   113
O.-P. Hilmola, *The Sulphur Cap in Maritime Supply Chains*,
https://doi.org/10.1007/978-3-319-98545-9_7

7.1   INTRODUCTION

In supply chains, product inventory and delays in the transportation process are often overlooked issues (Salam and Khan 2016). They are just a fact of the process, about which nothing can be done. However, they are a very important part of competition since corporations and outsourcing models have been progressing increasingly towards a global approach (e.g., Kumar et al. 2009; Holweg and Helo 2014). Therefore, components, semi-finished items and final products can come from anywhere, and overall products could travel around the world (not necessarily only once) before finally arriving with a customer or consumer to be used. For example, the raw materials might come from Australia and South-Africa, the whole machine intensive operations may be completed in Germany, and the assembly operation may be carried out in Ukraine. Assembly operations are fed by components that comefrom China and Vietnam. Then corporation has a global delivery center somewhere in Europe, where some testing, customization and packaging is completed (e.g., Jammernegg and Reiner 2007). Final delivery could take place in Japan or the USA. The world is indeed flat and hyper-competition is a reality and a part of everyday life, and therefore further specialization and global strategies are required (Buxey 2005; Puig et al. 2009; Kovach et al. 2015; Lorentz et al. 2016).

These complex supply chains and networks will face difficulties as sulphur regulation is implemented globally for the maritime sector in 2020. If supply chains slow down their speed (e.g., with ships), then the question arises as to whether costs relating to inventory holding and price erosion (lost money due to falling behind with up-to-date fashion, technology, and/or the physical freshness content of products; Grimm 1998; Helo 2004; Camejo et al. 2012) both increase. If time delays are considerable, supplied products could even be in danger of ending in liquidation sales (partially or completely), where prices could be 10–15% of the original retail price (Wood et al. 2005). This effect is rather large in supply chains as in fact the inventory holding curve is flat for the entire transportation process (Hilletofth et al. 2011), which is totally different from the 'sawn profile' of warehouse/retail store goods held in inventory (there is one bigger shipment every now and then, but continuous daily/weekly demand). In the transportation process, the used transportation lot size for goods means that someone is paying for this entire lot spending time in the transportation process for weeks or month(s). If this is compared with wholesale or retail inventory holding, the order lot size basically means that company is holding half of this amount in the inventory. If the inventory is critical to success in manufacturing operations of just-in-time and lean, it is even more important in global supply chains, and particularly in continental (or otherwise long-distance) transportation with considerable delays. Based on Bushuev et al. (2015), lot sizing decision and carbon emissions are taking their early steps, and there is no common agreement, how emissions should be dealt in ordering decisions (models to include, besides direct emissions, emissions from unsold products and fixed carbon costs).

In the following sub-chapters, three main simulation model findings from different contexts are presented. First, Sect. 7.2 demonstrates with a simulation why speed and very short sea shipping with trucks and semi-trailers was one solution to tackling the effects sulphur regulation effects in Northern Europe in 2015. Thereafter, the situation within continental transports in the face of the 2020 situation is analysed through the three most used supply chains between Europe and Asia in Sect. 7.3. The simulations end in Sect.7.4, which illustrates cost effects on different transportation modes for the diesel market, and also discusses maritime freight rates in detail. Finally, Sect. 7.5 features the conclusions.

7.2   Finnish Supply Chain Options to Central Europe

In this section the simulation model is used numerous times (Ivanova et al. 2006; Forio Simulate 2018a) in an extended form in order to understand why very short sea shipping combined with trucks and semi-trailers has been so popular in connecting Finland with Central Europe. This simulation model was originally used in long-distance continental transport chains (especially in researching how the railway landbridge between Asia and Europe could prosper), but it was modified and extended to European supply chains. For example, it is now possible for delivery lead times (for the entire transportation chain) to have a length of 0.5 weeks too. Together with this, CO_2 emissions from the route were taken into account by seeing the supply chain as a combination of two different transportation modes (unimodal is also possible, but then primary transportation mode has a 100% share). Emission coefficients originate from previous research, which combines two well-respected emission databases (Hilmola 2017).

In this section, it is assumed that the container being transported is filled with high-tech electronics, with a value of 50 USD per parcel (sales unit); one container could carry 1654 of these parcels (since space is the constraint). High-tech electronics units lose value constantly, by 0.4615% per week (27.0% p.a.). The inventory holding costs are estimated to be 0.2885% per week (16.1% p.a.).

To transport containers with very short sea shipping option from Finland to Estonia and further down to Central Europe, it is estimated that this activity would take 0.5 weeks in total. The container utilization rate (of high-tech parcels) is 90%, and the freight itself (all costs in the supply chain) has a price tag of 2800 USD. It is estimated that the distance to Central European customer facilities is 2600 km, and 95% of this distance is transported via truck, and 5% is devoted to the RoPax ship. CO_2 emission costs in the following are estimated to be 20 USD per ton, and the entire supply chain is liable for these costs. The container weight is set to 12 ton.

Since the supply chain has such a short lead time performance, the price erosion and inventory holding costs will not increase that significantly as compared with the pure freight price (Fig. 7.1). Together, these two account for around 279 USD, and the total cost in Fig. 7.1

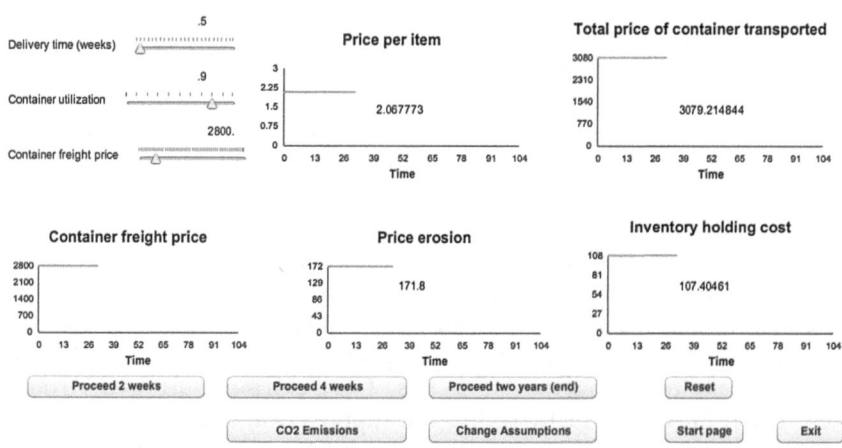

Fig. 7.1 Hypothetical supply chain costs from Finland using very short sea shipping Estonian (Tallinn) route with trucks and semi-trailers to reach Central Europe (*Source (simulation) Forio Simulate 2018a*)

is around 3079 USD. On the item basis, the overall cost is around 2.07 USD per high-tech parcel. CO_2 emissions could be seen as a weak part of this supply chain configuration—are somewhere above 3 tons. With a rather conservative CO_2 emission price, this corresponds to additional cost of 62 USD in the future (Fig. 7.2). Emissions are the weakness of this supply chain, but with conservative CO_2 emission costs, these truck-based supply chains will not diminish to nothing. The overall costs will increase only by 2.1%, if CO_2 emission costs are incurred.

The weakness of this supply chain solution could be changed to a real strength if an electrified railway option were to be available from Estonia (Tallinn) onwards to Central Europe (planned Rail Baltica investment). Here, the railway mode instead of the road in the hinterlands would make CO_2 emissions decline by 70% (depending on many factors, such as the extent to which trucks would still be used at feeding RoPax ships, in final mile delivery, and how transshipment functions are arranged). Of course, the total costs would be little bit higher, since the lead time of this option would not be 0.5 weeks, but merely one week. What would be lost in price erosion and inventory investment would partially be gained in lower CO_2 emission costs. However, this holds only if railway transportation has a similar freight price to road.

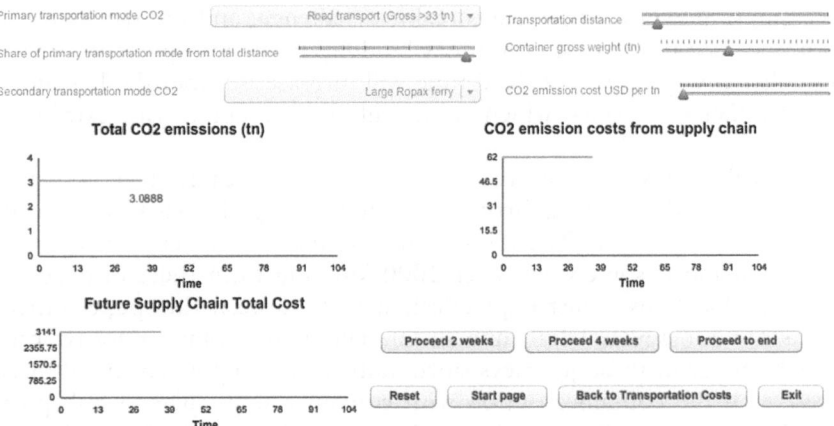

Fig. 7.2 Hypothetical supply chain CO_2 emissions from Finland using very short sea shipping Estonian (Tallinn) route with trucks and semi-trailers to reach Central Europe (*Source (simulation)* Forio Simulate 2018a)

To give some comparison and perspective on the Baltic Sea supply chains, another supply chain scenario was simulated with the assumption of mostly using container ships in the delivery. The lead time for the transportation service is estimated to be three weeks (from door-to-door). The overall freight costs of using container feeder ships and trucks (in hinterlands) would be somewhat lower, around 2400 USD. In the following is assumed that the supply chain has the same distance as before (2600 km), and 95% of transportation activity is accomplished with container ships. Weight, amount of parcels in container, price erosion, inventory holding costs and CO_2 emission costs remain as the same as before.

Even if, on the basis of freight price, this option is lucrative, it is actually really expensive if price erosion and inventory holding costs are taken into account. The total price of container being transported will reach above 4000 USD. On the item basis, it will increase to 2.74 USD. A long lead time has always been weakness of container shipping, and it will prevent some product groups from using this chain alternative. However, as its strength, this supply chain option has very low emission levels, and holds low possible emission costs (e.g. as compared with the earlier trucks with semi-trailers implemented RoPax supply chain) is approximately 60% lower. So, the future is looking bright for container ship-based supply chains, if only costs and CO_2 emissions are used as measures. Slowness of service is really a challenge, and in practice it will not only lead to increased inventory holdings in the transportation chain, but also at warehouses (as lead time will increase, so does the likelihood for product stock-outs, which could only be prevented with extra item holdings) (Fig. 7.3).

In reality, this simulation result is not that applicable since container shipping and distributing further to Central Europe by truck is not really a practical alternative. The distance to the main container hub sea ports from Finland is somewhere over 2000 km, and from there distribution to some locations is just impractical, such as Eastern Europe, Southern parts of Germany and Italy and so on. Therefore, distances are typically much longer in these journeys since, initially, the supply chain will head west with the container option with container ships (like to hub ports of Rotterdam or Bremen/Bremerhaven) and around Denmark, and from there to the east again, or southeast by hinterland transports. It was not only sulphur regulation and the advantage of speed (to significantly lower price erosion or inventory holding cost) that have caused the huge

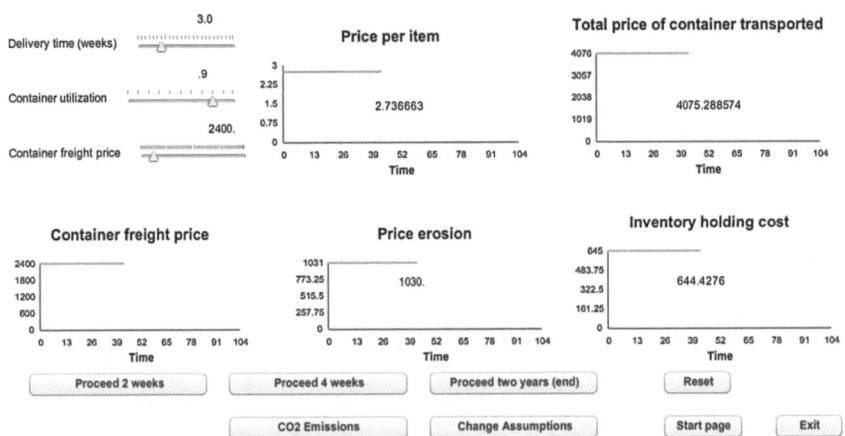

Fig. 7.3 Hypothetical supply chain costs from Finland using the container shipping route through the Baltic Sea, and hinterland transport with trucks (*Source (simulation)* Forio Simulate 2018a)

popularity of the Baltic route over the decades, but also overall costs and competitive distance to important and emerging European countries have played their role too.

In earlier research in the field, like Vanherle (2010, 2018), similar to the argument above, two European supply chains were examined where the short sea shipping option was compared with the supply chain dominated by the road transportation alternative (both cases of course contained both road and short sea shipping components, but neither one was dominating in the examined options). Two supply chain routes were compared (Vanherle and Delhaye 2010; Vanherle 2018): from Belgium (Kortrijk) to Norway (Oslo), and from France (Amiens) to Russia (Moscow). In both of the cases, the road-dominated option of course emitted more CO_2, but the difference in favour of the short sea shipping-dominated supply chain was not so significant (e.g., as it could be assumed a priori that container ships would result in very low emissions). Since research used real routes, and estimated emissions based on the specific transportation equipment used, short sea shipping resulted in around 30% lower CO_2 emissions. In Figs. 7.2 and 7.4, the difference was 60%, but in reality and practice, it is much lower. The reasons is simple: since road transportation-based supply chains are much shorter in

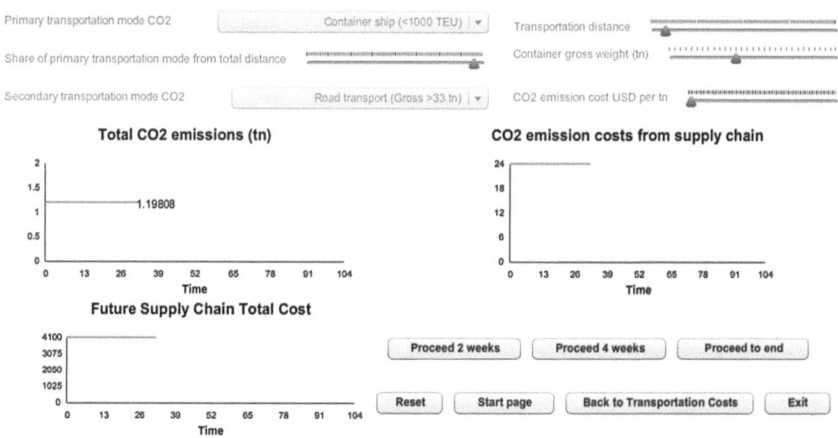

Fig. 7.4 Hypothetical supply chain CO_2 emissions from Finland using the container shipping route through the Baltic Sea, and hinterland transport with trucks (*Source (simulation)* Forio Simulate 2018a)

distance (more direct), and do not make detours, like container ships at the Baltic Sea, they do not polluting in that extreme fashion as argued based on emissions per km. In addition, notable in Vanherle (2018) research is the finding that short-sea shipping-dominated supply chains still emit more sulphur emissions at the Baltic and North Seas than road transportation-based chains. The reason is simple: road transportation used diesel fuel that contains only 0.001% of sulphur (as compared with a maximum of 0.1% at Baltic and North Sea for ships).

7.3   FINNISH SUPPLY CHAIN OPTIONS WITH ASIA

For the purposes of shedding light on the forthcoming global challenges of 2020, and the implementation of global sulphur regulation, it is important that a brief review of continental supply chains is made with the simulation tool. In this section, the three most common options are analysed: (1) using short lead time air freight, (2) using conventional and slow container shipping supply chain, and (3) using emerging new railway landbridge. The following analysis clearly shows that container shipping is currently the superior option, but its negative aspect is the same as before—loss of competitiveness in terms of price erosion and inventory holding cost. Air freight is not the answer in the long-term due to sustainability issues, and namely from the high CO_2 emissions produced.

The same high-tech product is analysed in the following, as it was earlier. The only change is weight of container, which is set to be a little bit lower, 10 ton. The distances between options do vary, and the container transportation network is the longest route, wheras flights and railway transport through hinterlands are the shortest. The distance using container shipping option is more than twice as long to Finland, for example.

For air freight-based supply chains, sulphur regulation will not cause any negative side-effects. Air transports uses tax-free fuel (kerosene), and its pricing is follows its own global patterns, which are typically tied to oil barrel prices. So, potentially in the short-term, air freight-based supply chains could be the beneficiary of changes of the year 2020. However, there are some caveats on the way.

The hypothetical supply chain uses air transport (starting from Hong Kong or Shanghai, for example) to reach Finland. Short-distance hinterland transport is completed with trucks at both ends. Transportation distance of the supply chain is estimated to be 8000 km, and 92% of this is accomplished via air. The lead time of activity is set to one week, and the air freight price is 5000 USD per container (although airlines are using own container system, not shipping containers). The cost of CO_2 emissions are set to 20 USD per ton.

Regarding the current costs of freight, price erosion and inventory holding costs, it is evident that with one week's lead time, the effect to latter cost items is marginal with respect to overall costs. The total price of a container being transported from Asia to Finland is 5558 USD (see Fig. 7.5). On the item basis, this corresponds to 3.73 USD. However, a challenge for air transport in the medium term will be emissions (also Arikan et al. 2014). This supply chain emits a staggering amount of 47.3 tons of CO_2. Even with conservative CO_2 emission payments, this will correspond to nearly 950 USD of additional payments (see Fig. 7.6) to this supply chain alternative. This is clearly a threat in a post-2020 world, where the European Union will start to implement its ambitious CO_2 emission reduction strategy.

In container shipping-based supply chains, transportation costs are not actually the problem. In Fig. 7.7, these are set to 2000 USD per container being transported. Total transportation distance in this chain

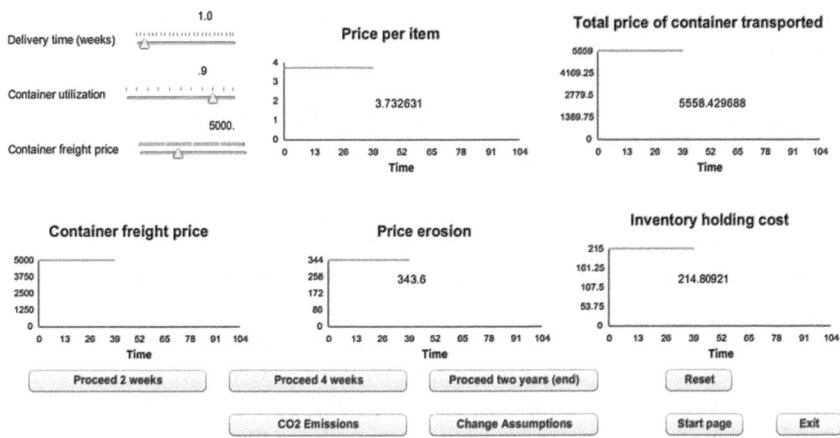

Fig. 7.5 Hypothetical supply chain costs to reach Finland from Asia (e.g., Hong Kong or Shanghai) using mostly air transport, and in hinterland truck (*Source (simulation)* Forio Simulate 2018a)

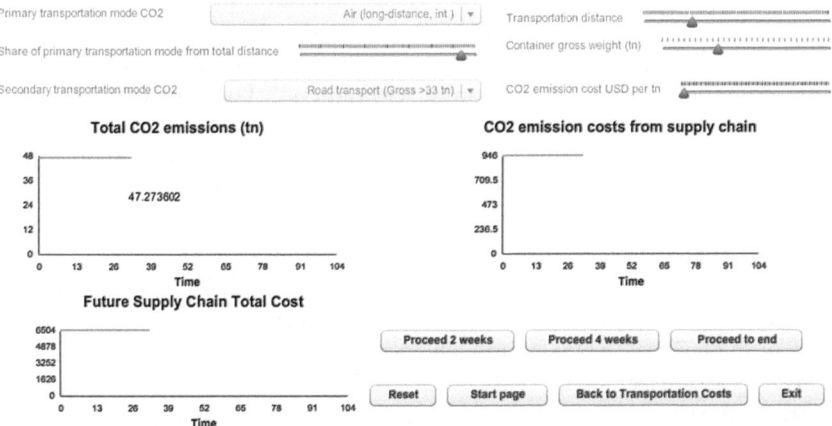

Fig. 7.6 Hypothetical supply chain CO_2 emissions to reach Finland from Asia (e.g., Hong Kong or Shanghai) using mostly air transport, and in hinterland truck (*Source (simulation)* Forio Simulate 2018a)

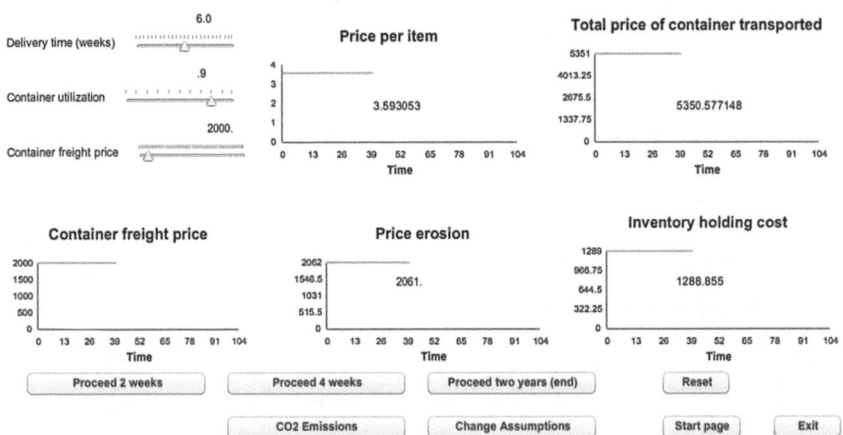

Fig. 7.7 Hypothetical supply chain costs to reach Finland from Asia (e.g., Hong Kong or Shanghai) using container ships (deep sea and feeder) (*Source (simulation)* Forio Simulate 2018a)

is 22,100 km, and it is estimated that this will take 6 weeks to be accomplished. The share of larger container ship (8000+TEU) from overall distance transported is 90%, and container weight is set to 10 ton.

Due to the considerable delay of transports, price erosion is nearly the same for this supply chain as it is for the transportation costs (Fig. 7.7). The inventory holding cost is above 1200 USD. These three factors together will result in a total price for the container being transported of 5350 USD. It is not that far from the cost for an air freight-based supply chain. On an item basis, the cost is 3.6 USD. Based on this, it is understandable that implementation of sulphur regulation will lead to some loss of volume in continental sea container transports since freight prices have to be increased. However, this could potentially lead to undesired environmental consequences since valuable products, with even some price erosion possibility, are more desirable to be transported with air freight. As Fig. 7.8 illustrates, container shipping-based supply chain in continental transport is very environmentally friendly, causing only around 3.4 tons of CO_2. Any additional costs from CO_2 emissions are small.

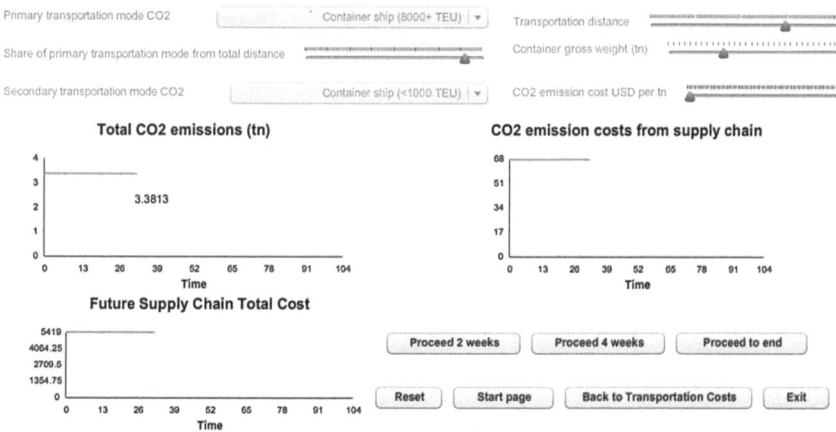

Fig. 7.8 Hypothetical supply chain CO_2 emissions to reach Finland from Asia (e.g., Hong Kong or Shanghai) using container ships (deep sea and feeder) (*Source (simulation)* Forio Simulate 2018a)

Using a hinterland-enabled railway landbridge has become increasingly more popular in recent years (Rodemann and Templar 2014; Yang and McCarthy 2013; Moon et al. 2015). This activity is taking place mostly between China and Central Europe. Earlier, in the late 1990s and early 2000s, rail (container) volumes between Asia and Finland were high (Hilmola and Lorentz 2012; Panova 2016). There are some signs that this activity is picking up again. One reason for this is the financial support given to railway transport through the Chinese 'one belt and one road' initiative. In addition, hinterland infrastructure has been improved across the entire route.

In the following, it is assumed that container transport from China to Finland, accompanied by the necessary road transport door-to-door will cost 3000 USD. The distance of the supply chain is 8900 km, and railways account for 95% of the distance being transported. Delivery lead time in this option is three weeks.

Regarding overall costs, the railway-based supply chain is attractive compared with the two earlier options. The total cost is somewhere below 4700 USD, and on an item basis, it is 3.14 USD. Of course, the freight cost is higher, but price erosion and inventory holding cost is half of the container shipping-based chain. With these hypothetical examples, it seems that the railway is most attractive in terms of gaining more cargo on its route, if the sulphur regulation of 2020 increases maritime transport costs considerably. For railways, the implementation of sulphur regulation could have some side-effects since in some parts of this land-bridge route, diesel traction is being used (especially in border areas of different countries). However, the effects are minor ones, and limited to availability and somewhat to the pricing of diesel fuel. It is notable that valuable products, but not those experiencing price erosion, are already attractive for this option. In Fig. 7.9, the total cost of freight price plus inventory holding cost is somewhere above container shipping-based supply chain (Fig. 7.7), but this could change to another direction in 2020.

In terms of its CO_2 emissions, the railway landbridge is environmentally friendly. In Fig. 7.10, it is assumed that a mixture of diesel/electric traction is used in the railway part of the supply chain, and in total CO_2 emissions are below three tons. This is due to low emitting railway transport, but also due to the competitive distance to reach Europe. An additional price tag or harm from CO_2 payments is rather insignificant for rail.

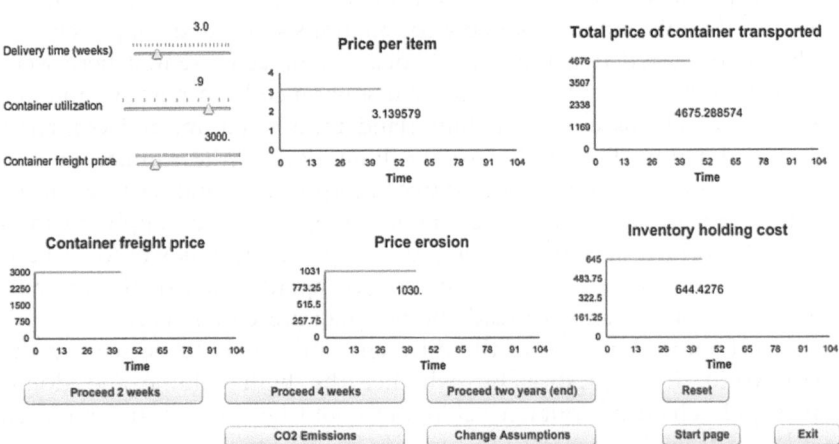

Fig. 7.9 Hypothetical supply chain costs to reach Finland from Asia (e.g., Xian) using mostly container trains, and in door-to-door transport trucks (*Source (simulation)* Forio Simulate 2018a)

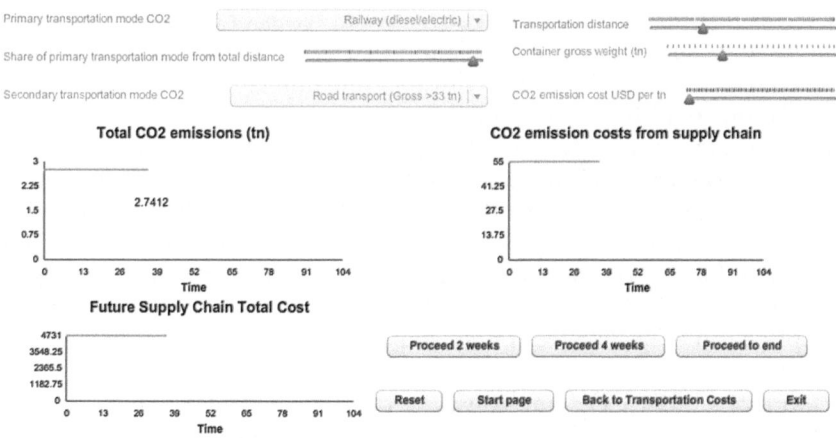

Fig. 7.10 Hypothetical supply chain CO$_2$ emissions to reach Finland from Asia (e.g., Xian) using mostly container trains, and in door-to-door transport trucks (*Source (simulation)* Forio Simulate 2018a)

So, from the continental perspective, container shipping will be hurt to some extent by the implementation of the 2020 sulphur regulation. Valuable products, even without the price erosion component, will be attractive for railways and possibly for air transport-based supply chains. It should be noted that the hypothetical examples presented here were not necessarily 'apple vs. apple' comparisons since both railway- and air-based supply chains contained hinterland transportation, and container shipping did not. This is due to the limitation of the simulation tool since it is only able to incorporate two transportation modes in environmental evaluations. In fact, a container shipping-based supply chain is also liable to include some hinterland transport, and this would make it a bit more expensive. Therefore, the cost competitiveness of air- and railway-based supply chains is already higher than presented here.

Even if some shift among the different options could be seen in the 2020 world, it is questionable whether the high volume and dominant supply chain of container shipping could be avoided and replaced at a larger scale. Transportation capacity has its limitations in two other options, and it could be the case that container shipping lines just do not transfer high fuel costs to customers, take short-term deficits, and adapt to the new situation. What is problematic in this scenario is the poor shape of shipping companies already (entire decade after 2009 crisis has been a poor time with declining freight rates). One countermeasure (Ronen 2011) is to further decrease the speed of container vessels, which would increase the fuel economy. This would, however, increase the negative impacts on the customer side (inventory holding costs and price erosion), and would also require reserve shipping capacity to exist. The container shipping market has spare capacity left due to the slump in 2009 and followed delivery of number of large-scale ships (with massive additional capacity) that this scenario is not that far from happening.

7.4   Different Scenarios of Implementation of Sulphur Regulation in Year 2020

Complexity and details of sulphur regulation change is difficult to comprehend. Therefore, for 2015, change at the Baltic Sea was built into a system dynamics simulation model, where both the diesel grade of shipping fuel changing and also in that time at the horizon being diesel market overheating (increasing oil prices) were incorporated in the same model (Hilmola 2015). Of course, it was evident in early 2015 that oil markets were so price depressed that the overall price was significantly declining and was offsetting the negative effects of sulphur regulation. However, elsewhere, shipping rates declined as they sustained their level in the Baltic Sea. So, change had its cost effects, but they were hidden under the oil price decline.

The simulation model contained originally the Finnish level of taxation for commercial shipping fuels (no tax), diesel trains (some taxes, but lower than typical), trucking (basically same tax level with consumers, but value added tax is deductible) and consumers using private cars (highest taxation level). In Finland, taxes on fuels are fixed EUR sums, which could be modified yearly (they are named energy content tax, carbon tax, and national security tax). The nly relative tax is value-added tax (VAT), which is paid by the final consumer, while railways and road transports may deduct this from their own sales VAT. Figure 7.11 and Forio Simulate (2018b) simulation model contains Finnish taxation on diesel fuels and VAT concerning the year 2018. It is of course much higher than in many European countries, but acts here as an example to illustrate how and why prices differ between different transportation modes.

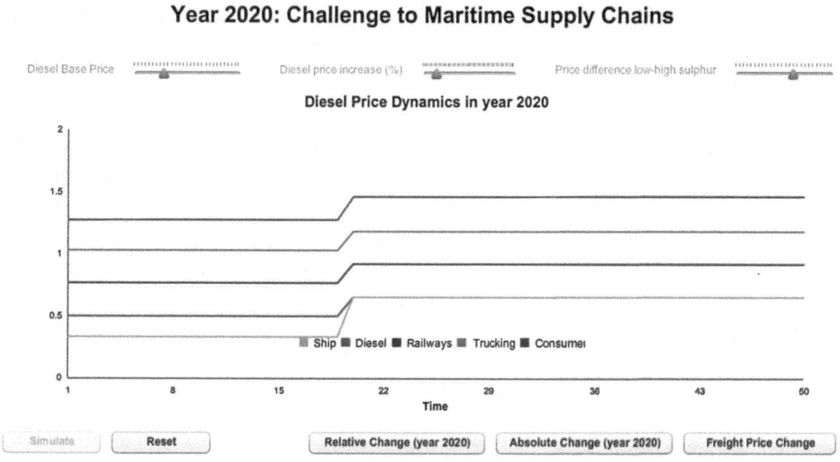

Year 2020: Challenge to Maritime Supply Chains

Fig. 7.11 Diesel price (absolute EUR price) changes in different usage groups with simulation default settings (diesel price increases in year 2020 by 30%, and ships need to use 50% more expensive 0.5% sulphur fuel) (*Source (simulation)* Forio Simulate 2018a, b)

The default settings of the simulation model set the price of diesel fuel (with refinery and distribution profits) at 0.5 EUR per liter. This is the level, which with taxes in April of 2018 in Finland, resulted in an average consumer price at diesel fuel stations. In the simulation model there is a 30% price increase taking place in the year 2020 (or 20 in the simulation model), and even if this increase is significant (Fig. 7.11), it results in moderate increases in diesel prices of trucking and consumer (due to heavy fixed tax load) since these diesel prices increase by 14.6% (Fig. 7.12). The effect on less taxed railway diesel is a 19.5% increase. However, the situation is worst in shipping diesel. Since it is a requirement from the year 2020 to use cleaner fuel (which is estimated to cost 1.5 times more), there is general price increase in the markets, and the shipping diesel shows huge 95% increase. This is the real danger in the implementation of sulphur regulation in 2020 globally—that simultaneously maritime supply chains are facing dearer diesel oil and also demands for cleaner and dearer diesel oil grade. It is not evident that this described event will take place, but its probability is rather high. As shipping globally starts to use lower-sulphur diesel, it might affect the demand of diesel production, and diesel prices could increase due to higher demand. It is

Year 2020: Challenge to Maritime Supply Chains

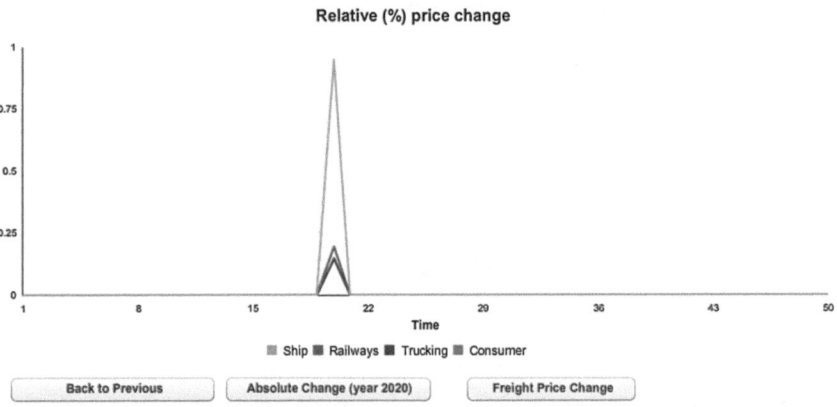

Fig. 7.12 Relative (%) diesel price changes in different usage groups with simulation default settings (diesel price increases in year 2020 by 30%, and ships need to use 50% more expensive 0.5% sulphur fuel) (*Source (simulation)* Forio Simulate 2018a, b)

known that diesel demand is rather inelastic (Hilmola 2015)—it is purchased in nearly the same quantities, even if prices increase.

In transportation logistics, it is typical that the cost of higher prices of fuel is passed on to customers. This is due to the low profit margin nature of the business, and also as practices have become accustomed to this over the decades (such as in shipping, with fuel surcharges). If the sort of situation that takes place in Fig. 7.11 is realized in practice, it would lead to rather sizable freight rate increases in maritime operations. Figure 7.13 shows the end result of this, which is a 28.5% freight rate increase. This is based on the assumption that diesel fuel represents 30% of overall shipping company costs. This of course in real-life varies between ship types, but it is anything from 10–50%.

By using the simulation and changing different parameter values (like price difference between low- and high-sulphur content maritime diesel as well as cost share of diesel in shipping), it is possible to make scenarios where the freight rates increase even 40 or 50%. This is of course the situation where the shipping fleet is not energy efficient (fuel plays significant role in the overall costs) and there is a high difference in the

Year 2020: Challenge to Maritime Supply Chains

Fig. 7.13 Diesel price changes and shipping freight level (diesel price increases in year 2020 by 30%, and ships need to use 50% more expensive maximum 0.5% sulphur fuel, diesel fuel costs of ship estimated to be 30% of overall costs) (*Source (simulation)* Forio Simulate 2018a, b)

prices of low- and high-sulphur diesel. It is possible that even without price increases in the overall diesel markets, but only changes internally in shipping markets (different grades of diesel oil), significant freight rate changes could still occur.

There is always a possibility that diesel prices in general in the world markets will decline heavily in 2020. This could also be taken into account within the model. Based on the default settings in the model, and just adjusting diesel price to world markets, it could be noted that change needs to be significant. It is somewhere around a 35% decline in prices that would offset price increases from the different grades of maritime sulphur diesel oil. With a decline of 50% in diesel prices, it is actually the case that prices in the shipping markets could even decrease by nearly 8%. It is of course a good question as to whether the year 2020 will repeat the situation of 2015 in the oil markets. Could it be this time different?

Different strategies are also easy to simulate with the built application. Some companies might choose to tackle the forthcoming change by a higher fuel efficiency. However, this might have effects on the capacity needed at sea (the number of ships); in addition, customer satisfaction in terms of longer lead times could soften the effects considerably. In some situations, the margin of the lower-sulphur grade is not that significant and diesel prices will not change at all; fuel saving could be the key to returning to the old freight price level.

Any of the simulation models presented here could be further developed as simulation models are available at applications (Forio Simulate 2018a, b) in Vensim format. This simulation program is free of charge at the evaluation level at Vensim webpages (Vensim 2018), and it relies entirely on well-established system dynamics simulation approach.

7.5   Conclusions

It is very difficult to forecast future events. In fact, the best actors in business prepare for different alternative futures, or in other words, scenarios of possible futures. Simulation is an ideal way to create such scenarios, and to see what the possible end results could be. However, often it is the case that scenario itself is not that interesting, but instead the means—why performance leads to such ends—are interesting. For example, in the earlier simulated different supply chains, and their cost performance was in wider perspective, and contained also CO_2 emissions. As a learning point, it could be concluded that freight cost is one important part of the competitiveness equation, but cost is also tied to speed. As sulphur regulation will increase the cost of maritime freight transportation, the importance of maintaining speed in the supply chain is stressed. If the relative competitiveness of maritime transport in the cost side will decline, it is rather logical that companies favour solutions, where this harm is minimized and speed is kept high. This is what has happened to unitized freight flows of Finland following the 2015 sulphur regulation change—short lead time solution to Estonia in maritime part with very short sea shipping style, and higher amounts of road transportation was the winning solution. Same development took place in Sweden—mostly through the sea port of Trelleborg to Central Europe. Of course, even old routes sustained the change, but they did not grow, and in some situations faced volume declines (like the Finnish–Swedish maritime route for unitized cargo). It could be argued that the 2020 sulphur regulation will cause similar kinds of effects. Some sea ports, shipping companies and routes will have a huge opportunity, but it should be built with overall package style, where the freight cost is the only one part of the supply chain (others packaged factors may incl. time, frequency, overall route length, amount of shipping operators and resource costs in the route).

It is vital that in the post-2020 era, new routes and alternatives base their competitiveness on promptness, frequency and speed.

As the continental simulation illustrated, air freight and railway-based supply chains will have their argumentation points and positive factors in 2020 change. However, air freight has a difficult position in the long-term as emission levels are so high. The railway solves many of the open issues of air freight supply chains, however, it is questionable as to whether these two alternatives will be able to take significant share from freight volume. They might experience long-term growth, for example 5–10% per year for a decade or two, but container shipping-based supply chains will innovate some other alternatives. One of them is that companies start to mix container shipping and air. In this mixture, lower CO_2 emissions could be achieved, as well as speed and promptness (Hilmola 2017). Another alternative is that container shipping companies start to think about their routes again—many of these are nowadays long ones, due to the current hub-and-spoke system. This may experience some changes in the forthcoming years.

In the last simulation model in this chapter, how diesel price changes and the sulphur regulation change in year the 2020 will affect the maritime sector and other transportation modes was analysed. The evaluation was made with Finnish taxation level and diesel prices in 2018 (early months); this country and its taxation system acted as a proxy for other countries. The simulation model gave learning points in the way that shipping diesel was previously very cheap, and if sulphur regulation change in 2020 takes place simultaneously to diesel price increases, it will lead to very significant undesired side-effects in terms of freight rates increases. Of course, the fuel economy of ships plays a role in this change and being economical could soften the transition somewhat. Other transportation modes, especially hinterland (road and railway), will increase their competitiveness, even in the situation of increasing diesel prices. They already have a fixed tax burden to carry, and the diesel market does not directly affect freight prices negatively as it does in shipping.

References

Arikan, E., Fichtinger, J., & Ries, J. M. (2014). Impact of transportation lead-time variability on the economic and environmental performance of inventory systems. *International Journal of Production Economics, 157,* 279–288.

Bushuev, M. A., Guiffrida, A., Jaber, M. Y., & Khan, M. (2015). A review of inventory lot sizing review papers. *Management Research Review, 38*(3), 283–298.

Buxey, G. (2005). Globalisation and manufacturing strategy in the TCF industry. *International Journal of Operations & Production Management, 25*(2), 100–113.

Camejo, R. R., McGarth, C., Herings, R., Meerding, W.-J., & Rutten, F. (2012). Antihypertensive drugs: A perspective on pharmaceutical price erosion and its impact on cost-effectiveness. *Value in Health, 15*, 381–388.

Forio Simulate. (2018a). Eurasian landbridge simulation application. Available at https://forio.com/simulate/olli-pekka.hilmola/eurasian-landbridge/simulation/. Retrieved 3 April 2018.

Forio Simulate. (2018b). Sulphur regulation & diesel price. Available at https://forio.com/simulate/olli-pekka.hilmola/sulphur-regulation-diesel-price/overview/. Retrieved 6 April 2018.

Grimm, B. T. (1998). Price indexes for selected semiconductors. *Survey of Current Business, 78*(2), 8–25.

Helo, P. (2004). Managing agility and productivity in the electronics industry. *Industrial Management and Data Systems, 104*(7), 567–577.

Hilletofth, P., Hilmola, O.-P., & Claesson, Frida. (2011). In-transit distribution strategy: Solution for European factory competitiveness? *Industrial Management and Data Systems, 111*(1), 20–40.

Hilmola, O.-P. (2015). Shipping sulphur regulation, freight transportation prices and diesel markets in the Baltic Sea region. *International Journal of Energy Sector Management, 9*(1), 120–132.

Hilmola, O.-P. (2017). Transport modes and intermodality. In Carlton, J. et al. (Eds.), *Encyclopedia of marine and offshore engineering*. New York: Wiley.

Hilmola, O.-P., & Lorentz, H. (2012). Confidence and supply chain disruptions: Insights into managerial decision-making from the perspective of policy. *Journal of Modelling in Management, 7*(3), 328–356.

Holweg, M., & Helo, P. (2014). Defining value chain architectures: Linking strategic value creation to operational supply chain design. *International Journal of Production Economics, 147*, 230–238.

Ivanova, O., Toikka, T., & Hilmola, O.-P. (2006). *Eurasian container transportation market: Current status and future development trends with consideration of different transportation modes* (Research Report 179). Lappeenranta, Finland: Lappeenranta University of Technology, Department of Industrial Engineering and Management.

Jammernegg, W., & Reiner, G. (2007). Performance improvement of supply chain process by coordinated inventory and capacity management. *International Journal of Production Economics, 108*(1–2), 183–190.

Kovach, J. J., Hora, M., Manikas, A., & Patel, P. C. (2015). Firm performance in dynamic environments: The role of operational slack and operational scope. *Journal of Operations Management, 37*, 1–12.

Kumar, S., Medina, J., & Nelson, M. T. (2009). Is the offshore outsourcing landscape for US manufacturers migrating away from China? *Supply Chain Management: An International Journal, 14*(5), 342–348.

Lorentz, H., Hilmola, O.-P., Malmsten, J., & Srai, J. S. (2016). Cluster analysis application for understanding SME manufacturing strategies. *Expert Systems with Applications, 66,* 176–188.

Moon, D. S., Kim, D. J., & Lee, E. K. (2015). A study on competitiveness of sea transport by comparing International Transport Routes between Korea and EU. *The Asian Journal of Shipping and Logistics, 31*(1), 1–20.

Panova, Y. (2016). *Public-private partnership investments in dry ports— Russian logistics markets and risks.* Lappeenranta University of Technology, LUT School of Business and Management, Industrial Engineering and Management, Acta Universitatis Lappeenrantaensis 689. Doctoral diss. Lappeenranta, Finland.

Puig, F., Helena, M., & Pervez, N. G. (2009). Globalization and its impact on operational decisions. *International Journal of Operations & Production Management, 29*(7), 692–719.

Ronen, D. (2011). The effect of oil price on containership speed and fleet size. *Journal of Operational Research Society, 62*(1), 211–216.

Rodemann, H., & Templar, S. (2014). The enablers and inhibitors of intermodal rail freight between Asia and Europe. *Journal of Rail Transport Planning & Management, 4*(3), 70–86.

Salam, M., & Khan, S. A. (2016). Simulation based decision support system for optimization: A case if Thai logistics service provider. *Industrial Management and Data Systems, 116*(2), 236–254.

Vanherle, K. (2018). Emissions race SSS vs. ROAD—Road versus Short Sea Shipping (SSS): Updating the 2008 comparison of emissions between modes. In *Proceedings of 7th Transport Research Arena TRA 2018* (April, pp. 16–19). Vienna, Austria.

Vanherle, K., & Delhaye, E. (2010). Road versus short sea shipping: Comparing emissions and external costs. In *Proceedings of International Association of Maritime Economists (IAME) conference*, Lisbon, Portugal.

Vensim. (2018). Free downloads—Vensim. Available at http://vensim.com/ free-download/. Retrieved 6 April 2018.

Wood, C. M., Alford, B. L., Jackson, R. W., & Gilley, O. W. (2005). Can retailers get higher prices for "end-of-life" inventory through online auctions? *Journal of Retailing, 81*(3), 181–190.

Yang, J., & McCarthy, P. (2013). Multi-modal transportation investment in Kazakhstan: Planning for trade and economic development in a post-soviet country. *Procedia-Social and Behavioral Sciences, 96*(6), 2105–2114.

CHAPTER 8

Conclusions

Abstract As the world turns further away from traditional fossil fuels, the business sector is taking significant steps into the great unknown. The forthcoming regulation change in sulphur levels in 2020 represents one of these challenges. Adaptation in Northern Europe to similar regulations in 2015 was rather mixed among shipping businesses and maritime-based supply chains. As an immediate response, shipping companies either started to use low-sulphur diesel oil or invested in scrubbers. However, as a long-term response, LNG is seen as promising new fuel, and new ships are already in use or being ordered by companies. For supply chains, this change in 2015 meant that traditional routes were reconsidered and so was the extent to which different transportation modes were used in these chains. It seems that road transportation was the winner in unitized transport and on the whole shorter shipping routes have become more favoured.

Keywords Sulphur regulation · Year 2020 · Year 2015 · Supply chains

Global supply chains are facing a significant challenge in the year 2020 as sulphur regulation in the world seas comes into effect. Many actors believe that this could be tackled with tactics, innovative strategy and state of the art technology. However, the situation with diesel oil and supply chains and transportation is similar to that of coal and heating as well as electricity production. Even if all indicators, regulations and

© The Author(s) 2019
O.-P. Hilmola, *The Sulphur Cap in Maritime Supply Chains*,
https://doi.org/10.1007/978-3-319-98545-9_8

technology development suggest abandoning oil and coal, it is still rather profitable and convenient for the users who to continue favour them. The reason is simple; these sources of energy have good daily availability, distribution systems, used equipment, and the sales as well as prices are well-known and existent. Using other than fossil fuels, and decision-maker shall take very risky step to the unknown. Think about the heating company trying to change its coal use to cleaner energy, like wood, gas or renewables. Changing to these would force the company to abandon part—or all—of its physical production and distribution assets. In addition, it is unknown and really difficult to forecast what the availability of the new source of energy will be, and its price. The more uncertainty at the business level, the higher the probability for failure. Ultimately, businesses, like heating company or shippers, are dependent on revenues, cash flow and profits. The same applies to companies using global supply chains—first on the agenda is sustaining its own operations. This is of course alongside fulfilling all of the legislative requirements of environment, society, employees, financial institutions and so on. It is most likely that tightening sulphur regulation will lead to variety of remedies and responses, and most of these are still based on old fossil fuel utilization at sea and hinterlands. A variety of responses is what the *Economist* (2018) has recently predicted, and it also further argued that smaller actors will be holding the losing cards. Environmental demands could also be seen as a barrier to entry to any industry, and the driver of further concentration of business in the hands of few. This is what also happened on a small scale within the northern Baltic Sea (analysed shipping companies).

This book aims to discuss the details of how North European countries adapted to sulphur regulation change in the Baltic Sea (and in part the North Sea) during 2015. It could be stated that in the short term, and overall, supply chains did not complete that great or strategical changes. However, many things changed in terms of details and at the tactical level. It is evident that the used routes, and the mixture of different transportation modes in the supply chains received actions, and together these resulted in changes within systems level. This is the first learning point for tackling 2020 challenges. At the moment, routes at sea are rather long, and, for example, the container shipping network is used to operating through its 'hub-and-spoke' system. Could and should these be challenged in the post-2020 world? Absolutely, as shipping becomes more expensive. However, even at a higher price, shipping is competitive, and in some cases the sole real alternative transportation mode, to access distant markets.

Its usage should just be reconsidered again in this new upcoming context. For example, in Europe, the great question is to what extent great 'hub sea ports' are needed in order to make shipping routes longer (as deep sea megaships visits hub ports, which are then served by different countries using feeder ships). Could the direct call to a nearby sea port be the answer (e.g., Tolli and Laving 2007)? Or should everything be designed and planned again by making supply chains shorter? By considering accessing Europe through some Mediterranean sea port, and then transporting items by road or rail to the final destination, instead of going through the hubs of Rotterdam, Hamburg or Bremerhaven. In some cases, this could make business sense. The same analogy applies to North America and using more west coast sea ports, instead of going through Panama Canal and transporting to the final destination in east coast. In China, this could mean that as the amount of sea ports serving trade are increasing, then the rather concentrated structure should be changed to become more versatile (in order to serve hinterland transport better, and make sea journeys shorter). Earlier research does not show that much opportunity in such developments (Notteboom et al. 2017); however, shipping being higher priced could change decades-old concentration models.

In Northern Europe, it was the case that shipping companies started to mostly use very low sulphur content diesel oil, and only in limited amount of cases did shipping companies invest in scrubbers. Olaniyi's (2017) research shows that most activities of installing scrubbers to ships were within one Danish-based RoRo/RoPax shipper, together with one Finnish originating (and also analysed in this book) company. However, publicly, this Danish company has stated that in the long-term LNG would be the solution within the strict 0.1% sulphur emission control areas (Ship & Bunker 2016), especially in newly built ships. All sorts of small improvements and approaches to increase the fuel economy were reported in the annual reports. Simply lowering sulphur diesel oil was quite a good decision in 2015 environment, when oil prices were declining. This gave shipping companies breathing space, and one company used this wisely to invest gradually in scrubbers for ships. This company experienced extremely successful financial years in 2015–2017 too. In the face of the 2020 challenge, it is said that for larger container shippers, MSC (Mediterranean Shipping Company) is applying the scrubber installation approach to tackle sulphur cap (*Economist* 2018). However, it should be remembered that shipping companies in Northern Europe have not invested that much in recent years, and asset amounts are declining. In addition, revenues are on slight decline (or in other words, clearly maturing). Thus, many investments are now at shipyards to be completed and delivered for future use.

Globally implemented sulphur regulation could be different from the implementation of sulphur regulation in 2015. The threat of supply constraints in low sulphur content diesel oil for ships could be real, and this in turn could have negative implications for prices (George 2018). It could mean higher prices and a wider spread between high sulphur content diesel oil and low grade. In this environment, there is a real danger that some shipping companies and supply chains use time as the competitive weapon, if they have scrubbers installed and if high content sulphur diesel oil is low priced. It could lead to the situation where CO_2 emissions increase since it is so cheap to transport in faster motion, and to differentiate from competitors (e.g., as argued in Lindstad et al. 2017) and eventually take the market share. Faster operations also release more capacity from existing fleet.

Book also showed that trucking and the wider use of semi-trailers was the answer to many supply chains to tackle the 2015 challenge. This is of course bad news for the environment, but it is reality and fact, and tells us something about the competitive situation of transportation modes. Road transports have been punished in recent years in Europe with different taxes, tightening of emission standards, and increasing fuel costs. Still, this was the partial answer to tackling sulphur regulation through very short sea shipping routes. It is an indication that trucking is still competitive underneath all the bad news and increasing demands. Perhaps shipping has been the 'free rider' in supply chains as its emissions standards and fuel taxation have been extremely low or non-existent, and this has created a situation where other transportation modes have not had that much space to compete against it in long-distance supply chains. Shipping is about to lose its competitiveness somewhat, and hinterland transportation modes will take their share. It would be ideal to have railways at the forefront of this change, but this book has showed that road transportation is actually the current solution. This is of course not sustainable in the long-term, but it is a short-term business fix for the supply chain challenge.

In the long term, it seems that the future of Northern Europe is at least partly in favour of using LNG in ships. In this book, it was found that many shipping companies have already ordered new fleet, and this did not only concern RoPax ships, but also raw material and container transport vessels. Increasing amounts of sea ports either already have existing LNG terminals in use or under construction (or they are in the design process). However, it is unknown how competitive LNG will be in the long-term. One uncertainty is still its fuel price. In this book, the fact that gas prices can be very low and competitive (North America), while they can also be high and uncompetitive (like in Japan) was discussed. Europe is still between these two. It is a rather interesting question as to what will happen to oil prices if LNG becomes mainstream. Could it mean that oil prices are not going to rise that much? Could they even decline? LNG is of course still taking its early steps, and there are many unknowns in the wider use of it. What are its real environmental effects (Anderson et al. 2015; Gilbert and Sovacool 2017)? Is it as clean as it is argued to be? Could it even result in increased CO_2 emissions (Gilbert and Sovacool 2017; Baresic et al. 2018)?

What is clear from the analyzed results of this book is the growth pains of tonnage volumes in the long-term in all the sea ports of Estonia, Finland and Sweden. In some cases, there is even a declining development in tonnage handling within the medium term (Estonia), whereas the others volumes are clearly maturing and experiencing slow growth (Finland and Sweden). In addition, in the survey many companies who reported the negative business effects of sulphur regulation were mostly from mass and high tonnage volume industries. Furthermore, most these companies were from Finland, which was geographically in the most disadvantageous position of all three countries regarding sulphur regulation change. This is just a note of caution that some industries need to be very careful and to invest time and energy in combating the 2020 challenge. Old and traditional industries have the most to lose. In the situation of the Baltic Sea, world trade and economic recovery in late 2016 and 2017 saved many heavy industry actors. The future will show how sustainable this recovery really was.

References

Anderson, M., Salo, K., & Fridell, E. (2015). Particle—And gaseous emissions from an LNG powered ship. *Environmental Science and Technology, 49*(20), 12568–12575.

Baresic, D., Smith, T., Raucci, C., Rehmatulla, N., Narula, K., & Rojon, I. (2018). *LNG as a marine fuel in the EU. Market, bunkering infrastructure investments and risks in the context of GHG reductions.* London: UMAS.

Economist. (2018, June 21). A wave of new environmental laws is scaring shipowners. *Economist.* Available at https://www.economist.com/business/2018/06/23/a-wave-of-new-environmental-laws-is-scaring-shipowners. Retrieved 27 June 2018.

George, L. (2018, April 11). Shipping fuel costs to spike 25 percent in 2020 on sulfuric cap: WoodMac. *Reuters, Business News.* Available at https://uk.reuters.com/article/us-shipping-fuel-costs/shipping-fuel-costs-to-spike-25-percent-in-2020-on-sulfuric-cap-woodmac-idUKKBN1HI1AT. Retrieved 16 April 2018.

Gilbert, A. Q., & Sovacool, B. K. (2017). US liquefied natural gas (LNG) exports: Boom or bust for the global climate? *Energy, 141,* 1671–1680.

Lindstad, H. E., Rehn, C. F., & Eskeland, G. S. (2017). Sulphur abatement globally in maritime shipping. *Transportation Research Part D, 57,* 303–313.

Notteboom, T. E., Parola, F., Satta, G., & Pallis, A. A. (2017). The relationship between port choice and terminal involvement of alliance members in container shipping. *Journal of Transport Geography, 64,* 158–173.

Olaniyi, E. O. (2017). Towards EU 2020: An outlook of SECA regulations implementation in the BSR. *Baltic Journal of European Studies, 7*(2), 182–207.

Ship & Bunker. (2016, March 9). DFDS: Scrubbers pay off for EVA compliance, but LNG bunkers more efficient solution for newbuilds. *Ship & Bunker, EMEA News.* Available at https://shipandbunker.com/news/emea/494964-dfds-scrubbers-pay-off-for-eca-compliance-but-lng-bunkers-more-efficient-solution-for-newbuilds. Retrieved 16 April 2018.

Tolli, A., & Laving, J. (2007). Container transport direct call—Logistic solution to container transport via Estonia. *Transport, 22*(4), 1–6.

INDEX

© The Editor(s) (if applicable) and The Author(s), under exclusive licence   141
to Springer Nature Switzerland AG 2019
O.-P. Hilmola, *The Sulphur Cap in Maritime Supply Chains*,
https://doi.org/10.1007/978-3-319-98545-9